Hunter and habitat in the
central Kalahari Desert

Hunter and habitat in the central Kalahari Desert

George B. Silberbauer

Department of Anthropology and Sociology
Monash University

Cambridge University Press

Cambridge
London   New York   New Rochelle
Melbourne   Sydney

Published by the Press Syndicate of the University of Cambridge
The Pitt Building, Trumpington Street, Cambridge CB2 1RP
32 East 57th Street, New York, NY 10022, USA
296 Beaconsfield Parade, Middle Park, Melbourne 3206, Australia

First Published 1981

Printed in the United States of America
Typeset by Progressive Typographers, Inc., Emigsville, Pa.
Printed and bound by Vail-Ballou Press, Inc., Binghamton, NY

Library of Congress Cataloging in Publication Data
Silberbauer, George B
Hunter and habitat in the central Kalahari Desert.
Bibliography: p.
Includes index.
1. G/wi (African People)   I. Title.
DT797.S48   305.8'961'06811   80-16768
ISBN 0 521 23578 2 hard covers
ISBN 0 521 28135 0 paperback

Contents

v

Preface

Much of social anthropological literature is about small-scale societies whose members have life-styles, world-views and philosophies
radically different from those of both the anthropologists and their
readers. As a discipline, anthropology fits squarely into the European intellectual tradition; its concepts and theories are consistent
with other areas of the philosophy and science of that tradition. It is
natural that we should seek to understand the unfamiliar in terms
that we already know, that we should try to explain alien cultures
through the use of concepts that can be reconciled with our understanding of our own way of life. This is, of course, only part of the
anthropologist's task. It is necessary for me to explain the socioecology of the G/wi Bushmen in a way that, I hope, makes sense not only
to those familiar with the European intellectual tradition but also to
those who have studied other societies and can coherently relate my
observations to their own specialized knowledge. The societies studied by anthropologists differ from one another to at least the extent
that, and in as many ways as, any one of them differs from an industrialized society. Anthropological theory, models, and concepts have
been developed from a universalistic perspective to bridge the differences among societies. Looking beyond the different usages, values,
and philosophies of small- and large-scale societies, we begin to discern the similarities among them and to realize that they have in
common certain themes in the ways in which they are organized and
in the sorts of problems faced by social man. The ways in which we
organize our different societies are variations on those themes, serving as different ways of facing what is a fairly uniform set of basic
problems. By considering social and cultural variety in the light of its
underlying similarity, we may come to a deeper understanding of
the extent of, and limits to, man's freedom to innovate.

It is not for anthropologists to say how that understanding shall be
used; that is the responsibility of men and women, common and un-
common, of the societies in which the anthropologists' ideas are read
and discussed. I am not putting forward the selfish, ethnocentric argument that social anthropological insight is the exclusive luxury of

industrialized societies, only to be used for their benefit. With our economic and political power, we who are members of those societies profoundly affect the lives of the less powerful. It would be well for us to have a clearer understanding of the nature of sociocultural systems in general so that when we encounter particular societies, as aiders, traders, or tourists, we may act with greater sensitivity and skill and do less inadvertent harm.

This book is about one aspect of the way in which one people lives. It is about the interrelationships of a population, its sociocultural system, and its habitat. The people are the G/wi Bushmen of the central Kalahari Desert in Botswana. The interrelationships are easier to see in the case of desert-dwelling hunter-gatherers than in a metropolis-oriented society like our own. The very much smaller size of the former population, the lesser complexity of the sociocultural system, and the comparative simplicity of the ecology of an arid biome like the central Kalahari mean that there are fewer factors to recognize and keep track of in conceptually linking them together. Furthermore, almost all the total stock of G/wi knowledge is current in the band as general knowledge (that is, as far as I could discover), and I therefore had access to a more nearly complete range of data than would be the case in a complex, plural society where much knowledge is esoteric and difficult of access. (This does not mean that I gathered a complete range of data – that would take a lifetime.)

The relations between industrialized society and its environment are obviously very different from those of a hunder-gatherer band and its habitat. The former are virtually global in their scope and are of immense scale and complexity. Transformations of energy, materials, and information within and between society and ecosystems are so many and so varied as to defy analysis by the simple means I have employed here. A direct comparison of the respective socioecologies is therefore likely to be sterile. To gain any useful insight into the relationships between a high-energy society and its environment from this study of the G/wi requires intermediate comparisons of socioecosystems of increasing scale and complexity. But at a simple level of comparison, it can be said that the members both of a G/wi band and of a large-scale complex society share the common lot of all life-forms in having to use the resources available in the environment to meet the pressures the environment exerts. Only by processing the material, energy, and information taken from the environment can any plant, human being, or other animal withstand the hazards of heat and cold, thirst and hunger, disease, predators, and so on sufficiently well and for long enough to raise its offspring

viii

to reproductive capacity and thereby prolong the existence of its species.

Our species, more than any other animal, has supplemented its genetically transmitted ability to utilize environmental resources to meet environmental pressures by developing extrasomatic, cultural means. Lacking effective claws to kill larger animals or teeth to rip anything but small and soft prey, our distant ancestors devised techniques of using sharp stones to enhance their predatory and defensive capabilities and invented missiles and ways of launching them with enough force and accuracy to compensate for bodily frailty and so on, through the inventory of man's gadgets and the repertory of their use. Such an inventory has allowed cultured man to make a larger and larger portion of the energy, materials, and information in the habitat available and gained for himself increasingly varied means of resisting environmental pressures. (All this cultural development required and stimulated anatomical and physiological changes. The emphasis here is on the role of culture and society, and for the sake of clarity, I have not discussed somatic changes.)

The cultural alteration of a species' relationship with its environment can be employed and function only in a social context. This is not to say that a spear needs a committee to use it but that the technology of its manufacture and use is too complex for each spearman to devise and develop *ab initio* for himself. Instead, the knowledge is shared among a cohesive, coherent group and is passed on to other individuals, groups, and generations. Our ancestors were able to develop extrasomatic, cultural aids because, at some stage and probably gradually, they largely abandoned instinct for learned behavior (or, if the sociobiologists prefer, they expanded the scope of instinct by introducing learned variations of its underlying themes). Whatever the case, the problem remains the same. Learned sequences of behavior, unlike instinctual programs, are not stereotyped and shared by the whole species. If one individual's learned action is to be effective, others must understand, hence learn, its meaning. Cooperation between people and coherent responses to one another's behavior entail the communication of, and agreement to, the intended meaning of action.

To learn the meaning of another's behavior does not necessitate the use of language: A homely example is provided by our chickens, ducks, and geese. They expect that when my wife appears with a bucket their feed troughs will soon be filled and they come rushing from every corner of the property. On occasion, however, the bucket is used not for feed but to collect eggs. In the latter instance the birds troop up to the house after my wife, angrily voicing their expectation of a feed, and will carry their protest meeting into the kitchen unless forcibly restrained at the door. Language makes it possible to define meaning more precisely than this and enhances the efficiency and versatility of communication. Through the use of nonconcrete, verbal symbols (instead of feed buckets) man gained partial freedom from the bonds of space and time by enabling those in the same speech community to communicate thoughts about the past and future and about things located elsewhere than in the vicinity of the conversation. It seems reasonable to suppose that this measure of four-dimensional freedom stimulated the human imagination to project states and actions to hypothetical, conditional circumstances so that people could build more elaborate patterns of interacton. Another domestic example illustrates this point: Dogs can be trained to obey commands. This is a simple pattern of interaction. Although the training is contractual in nature (reward in return for obedience), it is nevertheless one-sided, for the dogs have little capacity to initiate action in order to gain reward; that is the owner's prerogative. A more nearly symmetrical and complex relationship with one's dogs would result if the contract could be elaborated to include a clause specifying the rewards the dogs would get in return for killing the rabbits and foxes that plague the farmer. (Considering the jungle of labor relations, the farmer might be better to suffer the rabbits and foxes and leave his dogs as they are.) Without language, we are confined to a direct and immediate reciprocation and cannot elaborate a relationship to include more complex exchanges in a generalized mode or one in which a balanced exchange is conditional upon completion of agreed-on requirements over longer periods of time.

Language is preeminently learned behavior and therefore imposes the necessity for stable relationships to effectively communicate, and establish, agreement about its meaning. As I have said, it is also a means of broadening the potential scope and variety of relationships. It is also, of course, a means of expressing, or realizing, relationships and, furthermore, a means of regulating them. The structure of relationships impels, facilitates, and regulates the ways in which environmental resources are tapped, converted to use, and distributed among the personnel bound by the relationships. It follows that the state of the network of relationships is a factor limiting the variety and versatility of ways of using resources and of meeting environmental pressures (consider, again, the farmer and his rabbitting, fox-hunting dogs). The social organization of behavior is, therefore, a necessary (but not sufficient) requirement for the success of the human sociocultural strategy of survival. It is no less important an aspect of behavior than are the techniques by which people obtain their food.

As has so often been observed, there are neither free lunches nor free rides in nature. There are costs entailed in the establishment and maintenance of social order. Cooperation nearly always requires the participants to forgo some potential benefit in return for attaining a more widely shared goal. There are also opportunity costs; time and energy spent on decision making, for example, are not then also available for food getting. Learning, teaching, courtship, singing, and dancing are activities that have important social functions but produce no direct, concrete survival benefit. They are investments that, if all goes well, will yield a positive contribution to survival. Nevertheless, the investment must be fueled by food that is gathered or hunted or grown. In this sense the subsistence base sets a limit to social activity and its organizational tasks. The limitation is of the amounts of time and manpower that can be spared from food getting and says nothing of the nature of social behavior and its ordering nor of the choice that will be exercised among the permitted variety of ways of behaving.

Further organizational costs are incurred when, in adopting and developing a particular strategy, it becomes progressively more difficult for a society to switch to an alternative strategy. This is akin to the rule that specialization is achieved at the cost of evolutionary potential. A hierarchical society organized for authoritarian rule and backed by physical violence will not easily convert to an egalitarian style of decision making; a society in which the exchange of goods and services is an idiom of expression of kinship and other social relationships will find it difficult to adapt to a capitalist, profit-seeking socioeconomic system. Specific structural development tends not only to inhibit development in other directions but also to acquire a momentum of its own that constrains the society toward further commitment to the adopted strategy. The sort of thing I have in mind is what I describe later as the social hibernation of the G/wi band: When favorable conditions permit, the members of a band move as a group to a series of campsites throughout their territory; when the food supply deteriorates, each household disperses to its own isolated camp. This is not the only way in which the G/wi could organize themselves. Smaller bands, for instance, could manage with fewer shifts of campsite under the favorable conditions of midsummer to autumn and might even cope with the grueling first half of summer by living together as a united band but moving to fresh

Preface

campsites at shorter intervals. Assume, for the purpose of illustration, that such a strategy would secure the energy budget of, say, three average-sized households (15 people) as effectively as does the practice of social hibernation in a band of, say, a dozen households. The latter pattern confers the benefit of a fourfold increase in the total size of the band but is achieved at the cost of seasonal isolation of each constituent household. This means that each household must possess all the skills required for short-term survival on its own. It seems to me that this burden must tend to inhibit the development of specialized skills. Furthermore, the organization of the band must now be of such a nature as to attract members back into the joint mode of territorial occupation after the period of isolation but not so attractive as to inhibit their dispersal in the following autumn. I am not, of course, saying that the hypothetical alternative strategy of smaller bands would solve all problems; it would generate its own set of difficulties and organizational costs and tend to complicate, or even block, development in some other direction.

In explaining the relationship between population, sociocultural system, and habitat, the once accepted narrow and simple determinism that saw environment as causing culture lingers on in vulgar thought. It is perhaps understandable that this misapprehension should persist in a society as remote from the firsthand experience of the workings of habitat ecology as is ours (our view of population dynamics and sociocultural systems might also be somewhat clouded). After all, it is in the habitat that environmental pressures arise and it is from there that the resources for meeting the pressures are taken. As the environment sets the problem and also supplies the means of dealing with it, it is a seductive logical trap to conclude that the environment also determines behavior. The proposition, of course, is only true if there is but a single way of using the resources in a singular solution to the problem – conditions that seldom, if ever, obtain. Even something as specialized, mechanistic, and structurally simple as my pocket calculator allows me a choice of methods of doing my sums; the tripartite combination of a human population whose behavior is ordered by a sociocultural system in its relationships with its habitat will allow a much wider range of choice of ways and means of meeting problems. None of the component systems will be determinate in the narrow sense.

Possibilist models remind me of a wireless operator-gunner who whiled away long sea patrols by looking up "dirty words" in a large dictionary, which he took on every flight. He complained bitterly that, although he had searched every page in "H," he was unable to find "whore." Possibilism merely lists the range of things that could

xii

be done or that could happen, but says nothing of how or why and seldom covers the full range of why not; it explains very little about why some possibilities are exploited and others not.

A further fallacy in both determinist and possibilist models is their implication of a one-way relationship between a society and its habitat, that the society simply responds to what its habitat confronts it with. As I see it, it is, instead, a relationship of interdependence in that, for instance, change in the population will give rise to responses in both the sociocultural system and the habitat. Furthermore, the habitat is, to some extent, an artifact of the population acting within its sociocultural system (i.e., an artifact of the society). I am referring not only to the concrete conseqences of the society's behavior (the huts that are made by a hunter-gatherer band or the cities that industrialized societies build) but also to the way in which the society perceives and construes its habitat and the rest of the universe and, consequently, defines that part of the habitat's resources that may be used and the manner of its use.

A systems approach is well suited to the analysis of a people's socioecology. It conveniently permits examination of population, sociocultural system, and habitat, each being viewed both as a system in itself and as a component of the larger socioecosystem. Interaction and interdependencies among the three can be traced by following exchanges of materials, energy, and information. Operation of the whole and the component systems can be perceived and explained in terms of transformations of these and in terms of structural arrangements and rearrangements in the systems. It is a conceptual framework for the study of relationships within wholes, a strategy of searching for critical junctures in these relationships and a vehicle for explaining the consequences of interaction along the vectors of relationship. It is not a substitute for a theory to explain these consequences but a means whereby the differing areas of relevance of complementary theories can be reconciled and integrated. Incorporating, as it does, the concept of entropy, systems theory recognizes that the systems under examination need to be maintained if they are not to fall apart, cease to function in their accustomed manner and mode, and lose their identity.

As a social anthropologist, my main interest is the sociocultural system, and it is to this that I give most attention. If my descriptions of population or habitat stray beyond the limits of what their status as Durkheimian *faits sociaux* strictly justifies, it is because I needed to sketch them as coherent systems.

Some warnings must be posted. In what follows I interpret to you the relevant aspects of my experience of G/wi life. As I have said, my knowledge is not complete; absent material is a result of my failure to see and hear. Although our discipline developed out of innumerable studies of small-scale societies, we, the practitioners, nearly all come from metropolis-oriented societies and were reared or, at least, educated in large cities. Generally, social relationships in urban situations are notoriously fragmented, attenuated, and shallow, blunting our perception of others and of ourselves. I misgive our understanding of the deeper layers of interpersonal relationships and communication. The sensitivity of G/wi men and women (and there is a legion of other examples) to the behavior of their band fellows would be attributed, perhaps, to extrasensory perception if manifested in our milieu. (Some evidence suggests that there are Bushmen who possess unusual means of communication, but there is no need to invoke that argument here.) To live much of one's life in the same small community in the unprivate intimacy of a band and to have the lifelong habit of attending not only to *what* is said but also to a wide spectrum of indications of *how* it is said must sharpen perception to a degree that we, perhaps, cannot even imagine. The G/wi are not superhuman paragons of social virtue; if their skill is remarkable, it is a skill that anyone with comparable experience could develop. It is no more remarkable (nor less so!) than your ability to form a mental image of the G/wi from looking at the little marks on these pages. The G/wi customarily scan others' behavior so keenly as to read a cue as slight and fleeting as a momentary change in the set of a man's neck muscles, from which they know that he has struck a difficulty in the work he is doing, a work that demands the concentration of his attention to the exclusion of conversation around him. The consequent withdrawal into silence is explained, and potential frustration and annoyance are thus avoided.

Considering this matter from another angle, I am, of course, more fully informed of the circumstances of my children's lives than I am, for instance, aware of the conditions of my students' and colleagues' lives. However, although our family lives in a friendly, fairly small community, most of the knowledge I have of my children's interaction with their friends, teachers, and others is indirectly gained. How much fuller and richer would social life be if I saw for myself most of that interaction, knowing the other participants from a shared experience of most of their lives and able to discern their moods, thoughts, and reactions to what we share? My concern is that we anthropologists may be missing a substantial part of what we are examining, that the ephemeral, impersonal relationships that embrace most interaction in our large-scale native societies have left us with an ethnocentric impairment that prevents our perception of what is virtually an additional dimension of society. If this be so, if there is a network of communication too subtle for me to sense, then I have overlooked a large part of the information that guides G/wi behavior and am ignorant of the structures by which that information is coded. Had I knowledge of the texture, colors, and the intricate variation of pattern that band members weave into their social fabric, I might understand more clearly what I have discerned of the tapestry of their society.

Sociocultural systems, the objects of anthropological study, are extraordinarily complex. One might almost say that they contain more variables than one can count, let alone account for. In the holistic perspective of the discipline, any subset of variables is potentially relatable to any other subset, and the anthropologist has no certain means of knowing how long the chain of essential relevance might be. That is to say, one cannot be sure that a discovery made the next day will not refute results obtained up to that time. To compound doubt, what is learned of the operation of a sociocultural system is subject to an unavoidable and uncertain degree of selectivity and refraction induced by the idiosyncracies of personality and prior experience of both the fieldworker and his or her informants. The methods by which we conduct research can reduce the potential for error and omission but cannot protect us from every pitfall. Our findings are not unique; there are songs, poems, paintings, and novels that said it all long ago. But a songmaker or poet uses idioms that serve emotional needs – how such insight was achieved can seldom be explained. These idioms are also susceptible to differing interpretations. The statements of social scientists should be unequivocal in their meaning and should include the evidence on which they are based. The conclusions must be logically consistent with that and other known evidence and must be empirically refutable or verifiable. Research methodology is designed to help in attaining this ideal but, as I have said, we undertake the survey of a field of unknown dimensions and cannot be certain that the instruments we design and use will record all that it contains. The development of anthropological theory progressively reduces uncertainty, but theory building must be fed by research. That we are not in an impasse is, I believe, because there is a generous measure of redundancy in sociocultural systems and in research itself. What is overlooked today will eventually be found on another occasion or by another researcher. I also believe that anthropologists are attentive to intuition. We may not do it with the gifted elegance of the poets and songmakers, but we do take some account of gut-felt innuendo, and to be nudged by such nonempirical intimations is not necessarily inimical to scientific validity provided the methods of science are later used to test the results and to search for their explanation.

There is no one best way of carrying out a research project. In the same way that a fieldworker must recognize that intuition is a skill to be developed and used, he or she should work out a style of investigation that suits his or her individual inclinations, situation, and abilities. I studied the G/wi from a socioecological perspective because, as I explain in the introductory chapter, it made sense of what I saw. This book is a version of part of my doctoral thesis. When I first began to write up my work, my supervisor, John Blacking, now professor of social anthropology in The Queen's University, Belfast, encouraged me to persist with this line. In the chilly climate of British structural-functionalism, seemingly inimical to human life, this was dangerous heresy in the early sixties. I am grateful to John for his intellectual naughtiness and rejoice that he has remained an unrepentant freethinker. There have been many other socioecological studies of Bushmen (e.g., those of Richard Lee, Jiro Tanaka, John Yellen) and other hunter-gatherers (e.g., by Richard Gould and Nicholas Peterson on Aborigines), and some must wonder if this is the only way anthropologists see hunting peoples. There are, of course, other ways; there is the outstanding example of Lorna Marshall, honored tribal mother of Bushman scholars, whose approach was entirely orthodox, as was H-J Heinz's most competent analysis of the social organization of the !xõ: Bushmen. More recently, Alan Barnard built his Bushman research around kinship studies and Mathias Guenther concentrated on social change and religion. In a less conventional vein, Megan Biesele investigated !Kung folktales; to understand these in their context, she immersed herself in the life and thinking of the people. Each of these fieldworkers illuminated the general, as well as the particular, aspects of the sociocultural systems he or she studied. Whatever the emphasis in the investigation and point of entry into the system, each has achieved and communicated an understanding of the whole. None of us will understand all of the whole, and the variety of emphasis and approach that exists is necessary if we are to fill one another's gaps and clarify perception of sociocultural systems by showing them in the nuances and contrasts of our different perspectives. It seems to me that requisite variety is sure to follow if fieldworkers continue to develop their own styles of research.

Acknowledgments

It is sad that none of my G/wi informants is likely ever to read this book and appraise it; a teacher likes to see what a pupil has made of his lessons, and for his part, the pupil wants to hear what his mentors think of his efforts and to thank them for what they gave him. Perhaps children or grandchildren of some of my informants will read this; I ask them to remember my gratitude.

I thank Boy Magetse and Phuthego Matsetse for carrying a heavy burden cheerfully and competently. There are many others among my former Service colleagues whom I thank for their support, tolerance, advice, and instruction. Alec Campbell and his wife, Judy, gave me very special support and encouragement at times when things seemed bleak beyond endurance. Karl and Elise Weyhe, too, are true friends. The farmers of Ghanzi, even when we disagreed over something, were gracious hosts and never failed to invite us to *kom nader aan die huis*. They taught me much about many things. All of these people I thank for their good company.

Max Marwick and Des Cole gave me much of their time and knowledge and made available to me invaluable resources at the University of the Witwatersrand. They persuaded me to loftier ambitions than I started with and for that faith I sincerely thank them.

Richard Gould, Ed Wilmsen, and Sally White encouraged me to rewrite my doctoral thesis for publication and made detailed and helpful suggestions, for which I am very grateful. John Pfeiffer cast a spell on Cambridge University Press. I hope the spell has sustained that most patient of men, Walter Lippincott, in the difficulties I have caused him. However, one of his pleasures must have been the faultless typing of the manuscript by Mary-Lou Maroney and Joan Green of the Department of Anthropology and Sociology, Monash University. I thank them for their painstaking work, fitted into busy schedules and carried out amid constant interruption by me, my colleagues, and our students (to say nothing of that invention of the Devil, the telephone).

The ideas in this book come from more books than are cited and from more personal communications than are acknowledged. I learned much from those who accompanied us on journeys. It is impossible to keep account of the sources of knowledge and understanding. Although I have written this book and am responsible for the form in which the ideas are presented, it is very much an anthology of the thinking of my friends and colleagues in southern Africa, Australia, and the field of hunter-gatherer studies.

Long before writing my report on the Bushman Survey (Silberbauer, 1965), I discussed with my informants the recommendations I was formulating (the more important of these were submitted in a separate document) and explained, as clearly as I could, that I intended to publish an account of their life in the central Kalahari. They found it difficult to believe that a description of their ways would be met with anything other than scorn and derision. I reminded them of the people who had come out to visit them – of Alan Donald, government information officer, and his wife, Peggy; of Joe Podbrey, poet and editor of *The Mafeking Mail;* of Quill and Janet Hermans, gentle people who had responded to them with warmth and a sensitive appreciation of the logic of their ways; of my close friend Alec Campbell (now, fittingly, director of the National Museum of Botswana), who had helped me see more clearly the beauty of Africa. It was then that the G/wi could believe that there were people who could take a sympathetic interest in them, people who would not see them in the light of the national stereotype.

I have taken so long to finish this book that our children have come to regard it as some sort of elder sibling. They and my wife, Penny, will be heartily glad to see it leave home. I apologize to them for the nuisance my procrastination has been. As companions in the field, Penny and I look back on our time in the Kalahari with much happiness. As a qualified nurse and psychologist, Penny contributed more to our relationships with people and gave me more understanding of them than I can adequately acknowledge. Many of the ideas in this book are hers. Over the years she has striven mightily to cure my Germanic habit of contorted, overengineered writing (which is an occupational as well as an ethnic disease). Although the malady lingers on, I thank her for the great improvement her efforts have brought.

G. B. S.

Note on Orthography

G/wi is one of the click languages. The Roman alphabet is adequate for some languages that include clicks among their consonants (e.g., the Nguni Bantu languages). However, this would not serve the needs of G/wi. First, it has more click consonants than have those that use Roman characters. Second, G/wi has other phonemes to which the International Phonetic Association assigned the Roman characters that elsewhere designate clicks. I have therefore followed the IPA conventions with certain modifications to fit the normal English-language typewriter keyboard and the fonts common to most printers. I have also departed from IPA usage by positioning *g* and *n* before the click symbols to designate voicing and nasalization, respectively. This seems to me a more logical sequence of information and to occasion less aesthetic offense than the customary jumble of letters after the click symbol. From a purely phonetic point of view, there is nothing to choose between these two styles as the two characters are technically a digraph for a single sound, which is, in each case (i.e., nasalized, voiced, or unvoiced click), distinct from the others. My feelings may be biased by my earlier experience of Zulu and Xhosa as written languages in which the convention is to place the voicing (or nasalization) element first. If this is so, then I would point out that the practice is satisfactory in these and the other Bantu click languages, which have a firmly established tradition of literacy. Where an *h* represents aspiration of the consonant (whether click or non-click), I believe it should be postpositioned. This is in accord with the sequence of phonatory processes and also avoids making a typographical mess of the representation of aspirated-voiced and aspirated-nasalized click consonants. This, also, accords with Bantu practice.

I have omitted tone markings as I doubt that the information is of sufficient interest to most readers to justify cluttering the page with diacritics and thus raising the price of the book yet higher. I am fairly confident that two significant tonemes exist in G/wi, as stated by Koehler (1962). My earlier impression was that there were three tone levels, but I was unable to find the minimal triplets that could be ex-

Table 1. *Orthography of click sounds in the G/wi language*

	Dental	Alveolar	Palatal	Lateral
Voiceless	/	≠	!	//
Voiced	g/	g≠	g!	g//
Nasal	n/	n≠	n!	n//
Aspirated	/h	≠h	!h	//h

pected in a three-toneme language with a large proportion of mono-syllabic words. However, if there are only two tonemes, it must be recognized that there are also high and low ranges of these, that is, a high-high tone and a low-low tone, even if these are not phonemi-cally significant.

Vowel transcription is very broad but not, I think, likely to dan-gerously mislead anyone. A colon after a vowel indicates its ex-tended length and a tilda represents nasalization of the vowel over which it is placed.

The clicks are written as shown in Table 1. Two of these clicks, the unvoiced dental and lateral, occur in English. The first (/) is written "tsk-tsk" and means either "reproach" or, in South African English, "sympathy" (when it is usually followed by "aag, shaaame"). The lateral click (//) is understood by some horses as an instruction by the rider to break into a faster gait; its unmounted use expresses lascivi-ous appreciation of the female form. The other consonants have values approximating those for English excepting:

> *ph, th, kh* are aspirated, voiceless stops
> *kj* is a palatalized velar fricative
> *s* is a voiceless, alveopalatal fricative (like *sh* in shout)
> *z* is a voiced, alveopalatal fricative (like French *j* in *j'ai*)
> *x* is a voiceless velar fricative (like *ch* in Scots or German *Loch*)
> *kx* is a voiceless pharyngeal fricative
> *h* is voiced when it appears between vowels; after a consonant or a
> consonant cluster it indicates aspiration
> *w* and *j* are semivowels (as if the following vowel were preceded by
> *u* or *i*, respectively)
> *ng* is a velar nasal (like *ng* in song)
> ' is a glottal stop
> *r* is trilled, or is tapped (when it sounds like a *d*)

1

Introduction

The "ethnographic present" in this book is the period 1958–1966. After that, when the decade-long drought broke in the late sixties, there was a rush of Tswana and Kgalagari pastoralists and their herds into the central Kalahari. Previously held back by the long drought, desperate for new grazing because their stock had exhausted the pasture within reach of established wells and boreholes, the cattlemen took advantage of the good rains. With no adequate government control, they effectively dispossessed many Bushman bands of their territories. Exploration for, and exploitation of, minerals in parts of the central Kalahari will further disrupt the lives of the Bushmen. Development will undoubtedly bring material benefits, but Bushmen, at the bottom of the socioeconomic ladder, will gain least. Lately the Botswana government has moved to do something for Bushmen, but, for better or worse, the close-knit, self-sufficient organization of band society, which is described in this book, and the completeness of the band members' control of their society are gone. The "ethnographic present" is now the past.

The G/wi are one of many Bushman peoples of the Republic of Botswana (Figure 1). They live in the Ghanzi district and their country stretches from the northeastern corner of the ranching block down into the middle of the Central Kalahari Game Reserve and includes most of the western half of the reserve. In the 1960s, they numbered nearly three thousand, of whom rather more than half lived permanently in the central Kalahari as hunters and gatherers. The others lived on the Ghanzi cattle ranches as farm laborers or squatters or visited the ranches more or less regularly to take advantage of the water supply and avoid the thirst and hardship of winter and early summer in the desert.

They call themselves G/wikhwena (bush people; i.e., people of the thorn forests). To avoid the dilemma of either being awkwardly pedantic and distinguishing between G/wikhwesera (two G/wi women), G/wikhwema (one G/wi man), and G/wikxwisa (G/wi speech, the language) or perpetrating inaccuracies that would be the equivalent of "an Englishmen," I have taken the element G/wi as the

1

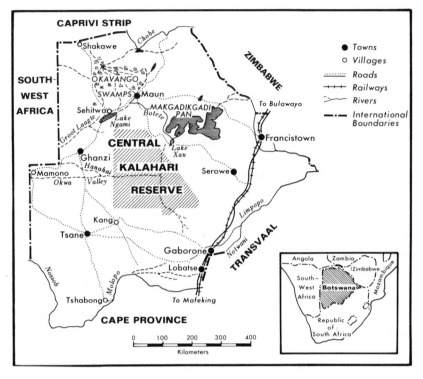

Figure 1. Republic of Botswana.

generic term for all things pertaining to the G/wikhwena. This parallels the ethnographic convention exemplified by referring to "the, a, or many Tswana" and "the Tswana language" instead of writing *Motswana, Batswana,* or *Setswana.*

The badge of language has social meaning and there is some feeling of unity among G/wi speakers. They are a timid people, fearful and initially shy and reserved in the presence of strangers. Such unity as exists within the language group is manifested by the reassurance and lessening of tension that is seen when a stranger is recognized as a fellow G/wi and by the subsequent eagerness to discover a link with him or her through mutual kin or friends. This is not to say that they will have nothing to do with non-G/wi; intermarriage with others occurs and is not stigmatized, and non-G/wi are accepted as friends, equals, or band fellows once acquaintance has overcome initial reserve.

It was once assumed that the contemporary Bushman population was the refugee remnant that had been driven into the present inhospitable habitat by the punitive raids of whites and Bantu-speaking Negroes. The assumption is refuted by historical records, which indicate that Bushmen tenaciously resisted to the last man in the areas where they are now extinct. Both Tswana tradition and the records of the earliest white travelers in the Kalahari region mention the presence of Bushmen and there is no evidence at all of refugee migration. Linguistic evidence indicates that the present distribution of Bushman peoples is long standing. First, the languages of surviving Bushmen do not match those of the areas where extermination occurred. Second, the gradation of languages and dialects among adjacent groups of Bushmen speaking languages of the same family is too orderly to be the result of the random influx of a refugee hodgepodge. Third, the nature and extent of the lending and borrowing of words to and from neighboring Bantu languages show that the present location of Bushmen is more ancient than the relatively recent period of genocidal conflict. Cultural adaptation to their present environment, including the existence of full inventories of names of significant fauna, flora, and other environmental features, is of a nature and extent quite inconsistent with refugee status. Finally, the Bushmen themselves have no tradition of retreat and migration into the Kalahari. It is, of course, quite possible that small numbers of refugees found their way north and attached themselves to, and became absorbed by, locally resident Bushman communities, but this is not the same thing as the large-scale displacement of populations claimed by the refugee myth.

There is some controversy over the name Bushman. Like other groups of hunter-gatherers (and those who hunted and gathered in historical times but have since adopted some other style of life), these people had no unifying organization that would have fostered a sense of shared identity and a common name with which to express it. Lumping them together in one category was the invention of outsiders, whether it was a Tswana calling them Masarwa or a seventeenth-century Dutchman referring to them as Bosjesmans, and the criteria of classification reflected the outsider's knowledge and his needs in so classifying them. These do not fit very comfortably the different needs of those who come later, and, with other criteria of classification, require a differently sorted set of categories. In the early 1970s there was a move in the scientific community to adopt the name "San" in place of "Bushman" because of the fancied perjorative connotation of the latter. As "San" already had "acquired a low

3

meaning" (Hahn, 1881:3) a century earlier, the choice could have been happier. "Bushman" distinguishes neither a language or even a family of languages, nor a unique style of living, nor a population that constitutes a particular physical type. Traill (1978) concluded that there are five separate language families among those of the yellow southern African hunter-gatherers. He found no satisfactory evidence of relationship among any of the grammatical structures of these language families and only rather low incidences of cognate forms (i.e., words with a discernible similarity of meaning and sound) between any two. Attempts at hypothesizing sets of rules governing sound shifts to account for the correspondences proved futile. It appears that there is not a family or even an order of Bushman languages. Their similarity is restricted to the use of clicks among their stock of consonants, which, because it is a feature unique to the speech of the hunter-gatherers and the Khoikhoi (the clicks in some Bantu languages – Nguni, Sotho, and some western languages – were borrowed from the Khoikhoi or the hunter-gatherers), tended to obscure the differences that closer examination makes obvious. One cannot distinguish hunter-gatherers from the Khoikhoi on linguistic grounds as one of the language families is common to both. G/wi, for instance, along with Nharo, G//ana, and others of Dorothea Bleek's Central Group (Bleek, 1956), is structurally and lexically akin to Nama and other languages of the Khoikhoi. Although these are referred to as Khoikhoi, or Hottentot, languages, there is no evidence that the G/wi and others acquired their languages from them. Because we do not know anything of the history of the language family, we label it Khoikhoi (or Hottentot) or Central Group Bushman, according to perspective.

Nurse and Jenkins (1977:16) conclude that "few morphological differences have been noted between the Khoi and the San except in overall size," that is, that Bushmen and Khoikhoi are the same physical type or of the same race.

In terms of economy, life-style, or sociocultural system, "Bushman" is again a nondistinctive term. Today there are Bushmen who are herders and there are Khoikhoi and Negroes who are hunters and gatherers. It seems, then, that Bushman is not a very exact term; it refers to an individual of the Khoisan physical type who follows a hunter-gatherer style of living, or whose ancestors were hunter-gatherers, and who speaks (or whose ancestors spoke) a language of one of the five families identified by Traill (1978). If this does not distinguish between a Khoikhoi who hunts and gathers for a living and a Bushman it is because the two are similar. Perhaps one should not expect labels of populations to have any very high degree of specific-

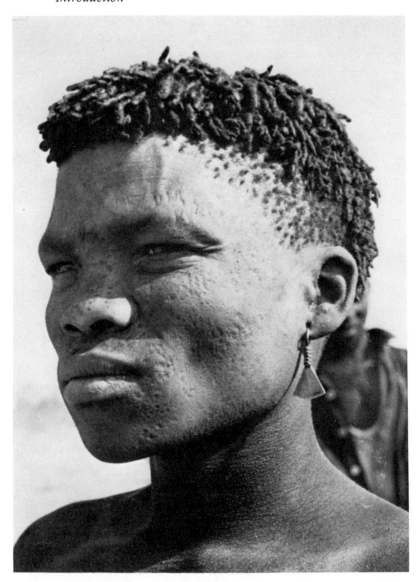

A G/wi man. Note the pockmarks, a legacy of the 1950 smallpox epidemic.

ity. After all, the term "Australian" may, with equal validity, be applied to a Turkish-speaking resident of Sydney, a member of the Pitjantjatjara people in the Musgrave Ranges in South Australia, and to Dame Edna Everidge of Moonee Ponds.

I shall use the term Bushman in reference to the G/wi because they

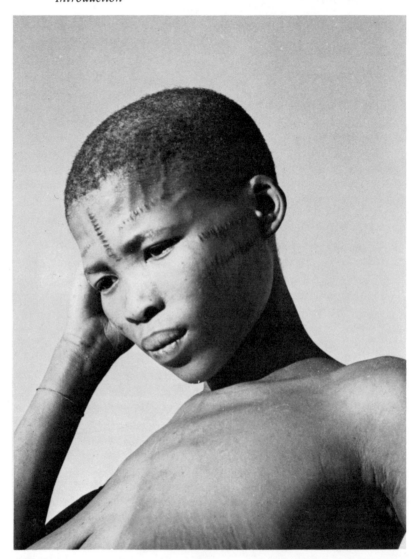

A G/wi woman. The marks on her face are recent cuts into which the ashes of medicinal plants were rubbed to cure an illness.

are included in the common (if imprecise) meaning of the word. Like other labels of other populations, this one may have to be hastily abandoned if the Bushmen unite and come to agreement in exercising their right to choose how they shall be known.

As Inskeep (1978) pointed out, we cannot be certain that the people who fashioned the artifacts of the Middle and Late Stone Ages in southern Africa were the ancestors of the Bushmen. There are still many archeological and paleontological gaps to be filled, but in the present state of knowledge it is reasonable to infer that they were probably ancestral to the population that, by 25,000 years ago, had come to closely resemble the modern Bushman. The putative ancestral type once inhabited the southern and eastern halves of Africa (Tobias, 1978). Although the antiquity of the Bushman presence in the Kalahari cannot be determined, there is no doubt that it has been the habitat of man for a very long time. Cohen (1974) described a Middle Stone Age site at Orapa diamond mine in central Botswana, to which he tentatively ascribed a date of 50,000 B.P. Similar sites are common along the eastern fringe of the Makgadikgadi and in parts of the Okwa valley, where I have also found Late Stone Age material.

As the wave of Negro migration spread across and down the continent from west Africa, the hunters and gatherers disappeared from the areas occupied by the black pastoralists and cultivators. Almost nothing is known of the history of their contact and it would be mischievous to equate black domination of the hunter-gatherers with either the ravages of white colonization or more modern instances of cruelties inflicted on Bushmen by their Bantu-speaking usurpers and overlords. Archeological evidence in the Transvaal Province of South Africa (R. Mason, pers. comm., 1973) indicates that there was a period of mutually rewarding and apparently peaceful coexistence of Iron Age Negroes and Bushmen in many localities. In these instances, dating from about 2000 B.P. onward, the local Bushmen were probably eventually absorbed into the larger population of newcomers, presumably the forebears of the present Sotho-Tswana peoples of central southern Africa. The rather later contact with Nguni peoples farther east seems to have been less amicable (Wright, 1971). Historical records of relations between Bushmen and Bantu-speaking peoples relate mainly to the Tswana and Kgalagari in and around what is now Botswana. Elsewhere, whatever relations may have existed between the two had been disturbed or destroyed by the combination of *Mfecane* (*Lifaqane* in Sotho) – the chain reaction of wars, pillage, and dispersal accompanying and following the rise of the Zulu – and the closely subsequent penetration and domination of the interior plateau by the Voortrekkers. John Campbell, who visited the southern Tswana in 1813 and 1820 describes Bushman raids on his hosts' cattle as if they were of fairly frequent occurrence. Retaliation ranged from giving the reivers a severe thrashing to the indiscriminate slaughter of men, women, or children encountered by the punitive party (Campbell, 1815, 1822). Missionaries and travelers who worked among and visited the Tswana in the second half of the last century report them as keeping Bushman slaves. Mackenzie, writing in 1870, said that the slaves were absolute property but that an ill-used Bushman might complain to the *kgotla* (chief's court) "and if his statement were borne out by fact he would get, not his freedom, but a change of masters at the discretion of the Chief" (Mackenzie, 1975:10). Southern Tswana tribes at that time treated their Bushmen better because, in Mackenzie's opinion, of "the proximity of the white man" (1975:10). He makes it clear that both the Tswana and the Kgalagari despised Bushmen and treated them very badly at times (Mackenzie, 1871:129–134). But few had much regard for Bushmen in those days; even the admirable Dr. Emil Holub refers to them in derogatory terms, despite owing his life to one (Holub, 1881, vol. 1:360 ff.). A. A. Anderson (1888:212–215), elsewhere sympathetic to the Tswana, described how hunting parties tortured Bushmen whom they suspected of having ostrich feathers or anything else worth taking. He was moved by these atrocities to petition the governor of the Cape Colony to extend Britain's protection to the Bushmen. Sir Henry Barkley refused on the grounds that the Kalahari was too distant. (Anderson's customary good humor failed him at this point and he wrote acidly of the governor's ignorance that the Kalahari was adjacent to his own colony.)

Hatred and hostility between Khoikhoi and Bushmen were reported by travelers from the earliest times, but whatever treatment the Khoikhoi and Bantu-speaking peoples may have dealt the Bushmen, the latter survived. When whites began to move inland from the Dutch East India Company's post at Table Bay, they followed the European pattern of closer settlement and more intensive use of land than was the habit of either Bantu or Khoikhoi. The result was the early development of murderous competition for hunting and grazing land. Initial contact with travelers was nearly always friendly, but when settlers moved in with their stock, the Bushmen, dispossessed of their hunting grounds, killed the cattle (which must have been tempting targets for hunters accustomed to wilder game). The settlers retaliated by killing the Bushmen and hostilities escalated to the point where each became the mortal enemy of the other, to be killed on sight. Their demise accelerated by epidemics of smallpox, measles, and other exotic diseases to which they had no resistance, the Bushmen were exterminated in a little more than 200 years after the start of the European thrust into the interior. To be more accurate, they seldom survived more than 50 years, or about two generations, of consolidated local settlement by whites. Both the Dutch and British governments at the Cape made a few halfhearted efforts to deflect the head-on collision of interests of settler and Bushman. These failed for lack of resources and determination, and both governments were forced by pressure from settlers to revert to open warfare. Even after the enactment of legislation recognizing the rights of Bushmen, slaughter continued under the guise of retaliation (see, for example, Wright, 1971:56). Many Dutch, German, and later British farmers were killed and many more ruined by Bushmen. But every Bushman band whose territory became a frontier farm lost its livelihood, living space, and the lives of its members.

In the remoter parts of southern Africa, settlement occurred in a later and rather more enlightened age. Men and women were not so driven by hardship that they were not able to devise a way of living in the country of hunter-gatherers without destroying them. Colonial rule was not so firmly established that it could guarantee the safety of lives and property, but it had a presence real enough to inspire some confidence of eventual retribution for murderers and thieves of whatever race. The country settled at about the turn of the century was poor land for ranchers, the competition they offered the hunter-gatherers was less intense, and in the arid interior each had something valuable to offer the other. Bushmen were the only help available to settlers accustomed to a labor-intensive style of animal husbandry, and the settlers, once they had sunk their wells, had a permanent and adequate supply of water to slake the thirst of every Bushman in the vicinity. At a time when educated Englishmen were poisoning waterholes in the interior of Australia and shooting manacled lines of Aboriginal men and women, barely literate Afrikaners had at least learned how to coexist with their Bushman labor. It was far from an idyll of racial and industrial relations, but it was not the genocide that had become habitual in the preceding two centuries.

In the early nineteenth century the G/wi and other Bushman peoples were the only permanent inhabitants of the country between the several Bantu-speaking tribes in the Lake Ngami–Okavango swamps complex and the Khoikhoi along the Black and White Nossob rivers. Except for the waterholes along the Ghanzi Ridge, this country has no permanent water and even the permanency of these waterholes is debatable. Although Galton (1889:166 ff.) reported good water at Rietfontein on the western part of the ridge in October 1851 (the height of the dry season), Andersson (1856:373), who had accompanied him, found far less water at the end of the wet season 18 months later. Baines (1864) and Chapman (1971), passing through in 1861, reported finding only small quantities of water at each of the waterholes. It seems likely that these holes would fill when there was a succession of good wet seasons and then gradually dry up during the following drier and drought years. Rainfall records for Ghanzi date only from 1923; since then, two very wet periods occurred (1934–1936 and 1973–1976), during which water flowed along what were previously regarded as minor fossil drainages, and pans that are normally seasonal remained full for several subsequent years. Local folklore recalls other earlier wet periods. Campbell and Child (1971) reviewed the evidence of ecological change in this and other parts of Botswana and were inclined to exclude significant climatic change as the cause of the disappearance of such species as elephant and rhinoceros from the Ghanzi Ridge. They attributed this to habitat change wrought by human interference. I have no argument with their conclusion but question whether these animals were, in fact, a permanent part of the fauna of this area. I suggest that these and other highly mobile species that require frequent access to water might have migrated into the Ghanzi area during the very wet periods and remained until the waterholes dried up. This is consistent with the contrasts recorded by the early travelers and with the sporadic movemements of fairly large herds of buffalo from Ngamiland to south of Ghanzi in the mid-1950s. The subsequent extension and fencing of ranches along the ridge and the construction of the game-proof fence along the Ghanzi-Ngamiland border have since put a stop to the migration of large mammals.

It is known that the Tawana of Ngamiland had cattle posts along the ridge, but there is some confusion and controversy about the permanency of these posts and the extent of Tawana suzerainty over the area. If, in fact, the climate of Ghanzi did include several wet periods in the nineteenth century, it is quite possible that the Tawana would imitate the migrating mammals and extend their pastoral activities into this good cattle country when the barrier of water shortage was removed by exceptional rains. The early travelers do not mention Tawana settlements, but their writings suggest that there was some Khoikhoi movement eastward onto the ridge and that there was a fairly substantial but irregular traffic in cattle and other commodities between the Khoikhoi and the Yei and Tawana of Ngamiland.

The Nharo, Tsao, ≠aba, and ≠xãũ//ei Bushmen, whose countries include the Ghanzi Ridge, were exposed to this sporadic intrusion and to the fairly regular passage of Khoikhoi, Bantu-speaking, and later white travelers. Relations between Bushmen and the Khoikhoi appear to have been as unfriendly in this part of southern Africa as they were elsewhere, but there was a measure of amity between them and the Tawana. Galton (1889:165–166) mentions the Bushmen's familiarity with both the language and country of the Tawana, and Major (later Lord) Lugard (1896) reported that the Ghanzi Bushmen rendered tribute to the Tawana chief. All the white travelers wrote of the friendliness and cooperation they received from these Bushmen, passages that are usually singular exceptions to their gloomy descriptions of their (the travelers') relations with other peoples whom they encountered in what are now Botswana and South West Africa (or Namibia). It says much for the manner of these Bushmen that they were able to win the confidence of men who, having suffered hostility from almost all whom they met and who were accustomed to retaliating with every means at their command, must have become habituated to extreme caution in their dealings with strangers. This early contact gave the Ghanzi Bushmen a foretaste of what was to come later and perhaps also the time to develop a measure of cultural resilience in the face of alien influence, which other more isolated Bushmen lacked.

Although G/wi country lies well to the east of the ridge and its traffic of travelers and short-term visitors, the G/wi would certainly have been involved in peripheral trade through the exchange network that stretches across the Kalahari. The first substantial white settlement was established at Ghanzi Pan in 1874 by Hendrik van Zyl (see Silberbauer, 1965:114 ff.). In the next two decades several more Europeans installed themselves at various waterholes along the ridge and in the Okwa valley south of Ghanzi, all well outside G/wi country. In the 1890s, however, Cecil Rhodes, prime minister of the Cape Colony, settled 18 Afrikaner families (mostly from the defunct Goshen Republic in the north of what is now the Cape Province) around Ghanzi in order to block a feared eastward expansion of the then German colony of South West Africa. By the turn of the century there were 37 farms, the easternmost of which was on the edge of G/wi country. The Bushman bands whose territories had been handed over to the ranchers became cowherds and farm laborers. The impact of the ranching economy was lighter than might, perhaps, have been expected. Each ranch was a small-scale operation. The whites were poor people with small herds and simple technology. Properties were not fenced, the nearest market lay nearly 1000 km across the desert, and stock could be trekked out for sale only when good rains fell and filled the pans along the route. The heavily fleshed, quality stock that southern cattlemen bred from imported blood could not stand the rough conditions and the long treks; consequently, the cattle the Ghanzilanders raised were small and scrubby and generally fetched low prices when they eventually reached the southern markets. The lives of these pioneer ranchers were nearly as hazardous as those of their Bushmen, and for some families the standard of living was not much different either. There is a tradition among the descendants of the original families (not observed by all but still staunchly followed by some) of honoring their obligation to the Bushmen. When drought overtook the advance party that had come to ascertain the suitability of the land offered to them by Rhodes, the men had to leave their wives and children in the care of local Bushmen while they made a dash across the Kalahari to get help. The drought worsened and it was nearly two years before they could return. When, at last, they managed to get back to Ghanzi, they found that, despite the severity of the drought, the Bushmen had taken excellent care of the women and children. The few remaining descendants of members of this party honor their debt by not refusing a Bushman's plea for food, shelter, or living space on their properties. Apart from the financial cost to occasionally hard-pressed families, the practice has also brought criticism from some newcomers who regard it as a nonsensical, sentimental nuisance to neighbors.

The government of what was then the Bechuanaland Protectorate was committed to a policy of indirect rule, of maintaining intact the political, judicial, and social organization of the dominant Tswana tribes (see Lord Hailey, 1953), and, as far as was practicable, of furthering the incorporation of non-Tswana, Bantu-speaking peoples into these tribes. Britain's protection had been requested by the Tswana chiefs, and the government saw its principal relationship and concern as being with them and their people. Gold had attracted miners to the Tate area in the northeast and there was an enclave of European cattle ranchers in the southeast. These white communities, which had some influence and were relatively wealthy, were regarded by the government more as an asset than a nuisance. But the Ghanzilanders, having served Rhodes's purpose, were virtually ignored. Poor, isolated, and of necessity inward-looking, they made a favorable impression on few of the district commissioners who were posted to this arid, lonely, and generally unpopular station. In conformity with the English South African stereotype current until the 1940s (see de Klerk, 1975:ch. 5), the Afrikaans they spoke was a badge of no-hope, poor-white status, a view made explicit in the reports and comments of many of these officials written in the first half of this century. District commissioners were, in principle, sympathetic to the Bushmen in their bailiwick but seldom had the opportunity or means of reaching an understanding of their way of life and its problems. They variously saw them as laborers ruthlessly ex-

ploited by ranchers, as occasional (or inveterate) stock thieves, or as
rather mysterious small people living in the desert. In the eyes of the
Ghanzilanders, the government spent its time damning them and
making pious statements of intent about Bushmen. But actually
doing anything about feeding the hungry, curing the sick, and providing employment was their, the ranchers', thankless task. Both
sets of views touched on the truth. Because of complaints made by
British, Australian, and New Zealand settlers (who, as English
speakers, may have been the only spokesmen available to the other
ranchers), the Ghanzi situation became polarized and the district
commissioners of the twenties and thirties were seen (and perhaps
saw themselves) as being either pro-rancher or pro-Bushman. Cap-
tain J. W. Potts, famous for his relief-work scheme to assist impover-
ished ranchers during the Great Depression, sternly warned his successor in Ghanzi, W. H. Cairns, when the latter announced his
intention to establish a Bushman reserve at Olifantskloof. It was the
Government's business, Potts said, to protect the public, and "if
your line of policy is adopted and a Utopian Bushman Reserve
created, you will find stock theft as prevalent as ever." Cairns perse-
vered, pointing out that public protection in the form of jail sen-
tences or lashes for offenders was not effective as these methods were
no deterrent to a man whose wife and children were about to die of
starvation. He started his reserve in September 1937 under the charge
of Police Sergeant de Lorme; it attracted a fair number of Bushmen
who made a living from trapping, selling skins and pelts, and working on the roads. Cairns faced bitter criticism at a stormy meeting of
ranchers at the end of that year. Stock theft had continued unabated,
the ranchers complained, and the reserve was useless. Nettled beyond endurance, Cairns turned on one rancher (a British settler who
eventually amassed a fortune) and pointed out that all but one of the
cases of stock theft had been committed by the settler's laborers;
their "wages" totaled a measure of about 1 kg of maize per week and
nothing else beyond occasional, small amounts of tobacco and an annual issue of secondhand clothing. His cows were all dry, so there
was not even milk for them to drink. The Englishman's retort, "if I
give them more, they just give it away to their relatives," seems to
have lost him the support of the others at the meeting who, it transpired, saw it as their duty (however onerous and unwanted) to feed
their laborers' kin in times of drought and famine. (Although it followed on a series of wet years, 1937 had very little rain and there was
a poor crop of esculent plants.) Cairns lost his temper when two An-
tipodeans suggested that the practice of a former policeman be re-

vived, namely, making Bushman prisoners run up and down in the heat of the day between the jail and the office (about 0.5 km) "to make them good and afraid of the Camp" (i.e., of government authority). He refused to "discuss such inhumanity," told them they could convey their wishes to the resident commissioner if they chose, and left the meeting. He was transferred to another station two years later and the Olifantskloof reserve was abandoned.

The same ambiguity and indecisiveness marked policy concerning Bushmen in other parts of the protectorate. When Britain assumed control of the whole of Bechuanaland in 1885, the institution of serfdom was ignored. Nor was official notice taken of it in 1895, when the country north of the Molopo River (i.e., what is now Botswana) became the separate Bechuanaland Protectorate. It was only in 1926 that Bushman serfs received official attention. As a result of the Ratshosa case, in which the return of two Bushman servants was sought by the wife of one of the Ratshosas, Lord Athlone, the high commissioner, addressed the *kgotla* of the Ngwato in Serowe and declared that government would not countenance any tribe's demand of compulsory service from another and that any Bushman who wished to leave his master's service was at liberty to do so (Benson, 1960:55–56). He went on to say that it was the duty of the Tribal Administration to make this clear to Bushmen and that he expected missionaries and government officials to encourage adherence to his declaration. Little or nothing appears to have been done until 1931 when Major E. S. B. Tagart was appointed commissioner to inquire into both the conditions of employment of Bushmen among the Ngwato and the system under which corporal punishment was inflicted on them (High Commissioner's Proclamation of July 11, 1931).

Tagart concluded that "the conditions under which the Masarwa [i.e., Bushmen] are employed by the Bamangwato tribe, while not as a rule involving excessive hardship, are sufficiently unsatisfactory to call for further investigation and action with a view to improvement" (Tagart, 1931:11) and recommended that a census be taken of Bushmen and that the names of Ngwato masters be recorded. He also recommended that the Ngwato set aside land for Bushman occupation, or if the tribe was not willing to do so, that the government reserve an area of crown land for the purpose. He further recommended "appointing native demonstrators in agriculture to encourage and assist in the cultivation of crops by any Masarwa who may elect to settle in village communities" (Tagart, 1931:16).

In 1935 the South African District of the London Missionary Society conducted its own inquiry into the circumstances of the Bushmen. The committee's recommendations (*The Masarwa*, 1935:26 ff.) leaned in favor of the integration of Bushmen into the Ngwato tribal structure and their assimilation to the Ngwato way of life, rather than in favor of the segregation that Tagart had suggested. Most of the other recommendations reiterated or enlarged upon those that Tagart had made.

As a consequence of Tagart's report, *The Native Labourers' (Protection) Proclamation, 1936* (No. 14) and *The Affirmation of the Abolition of Slavery Proclamation, 1936* (No. 15) were gazetted and Lord Athlone's declaration was read in every village and its implications explained. The Bushmen themselves rejected the idea of a separate reserve and stated their reluctance to leave their accustomed home areas. In mid-1934 a district officer, J. W. Joyce, was detailed to conduct the census that Tagart had called for and three years later submitted his report (Resident Commissioner's letter No. 6637/9 [II] of December 13, 1937). He reported having counted 9,505 Bushmen in the Bamangwato and he included his own set of recommendations. These were similar to those of the District Committee; government policy should aim at the assimilation and integration of Bushmen as full members of the tribe. In this he recognized that Bushmen would have obligations, as well as rights, if they were to become tribesmen.

In the meantime J. W. Potts, now resident magistrate of Ngamiland, had been conducting a Special Court at Maun to free Bushman "slaves." Sergeant Fox of the Bechuanaland Protectorate Police organized a series of rehabilitation schemes under Potts's supervision. These included teaching the cultivation of crops and the marketing of tanned hides and pelts. Fox appears to have spent a good deal of time and energy on this work and showed considerable enterprise, often paying for tools, materials, and seed out of his own pocket. He went about the villages explaining "the illegality of slavery" and protecting those Bushmen who wished to leave their masters. He evidently listened carefully to what the Bushmen told him and heeded their requests and complaints, for he reported taking no action where Bushmen wished to remain with their masters and detailed the difficulties of independence of those who had been released from servitude. Potts gave him full support and even persuaded the government to reimburse him for some of his purchases. The London Missionary Society evidently annoyed Potts, for he was scathing in his criticism of their "failure to come up with any useful, practical advice or help" and protested against their "innuendo that slaves are merely released and nothing done to help" (Botswana Archives, S. 440/4, 1935).

In January 1938 Joyce started a Bushman training community at

Lotlhekane where there were about 500 Bushmen, of whom about one-third had some knowledge of growing crops and had tried to cultivate their own. He took with him Gilbert Molaba, an agricultural demonstrator. Molaba had been seconded from the South African Agriculture Department on the recommendation of the pasture expert H. A. Melle, who had acquired some knowledge of Bushmen and their problems as a young cavalry officer in the South West Africa campaign of 1914 in the western Kalahari. Joyce and Molaba made a good team and the settlement did well. The district commissioner was not sympathetic, however, and seemed to go out of his way to frustrate Joyce's efforts. To add to the natural difficulties, a rather unconventional European who had "gone bush" and lived in the vicinity of the settlement had imposed himself on the local Bushmen as their headman. Apparently resentful of the independence that the scheme was bringing them, he also made life difficult for Joyce and Molaba. War broke out in 1939 and many government officials left to join the armed services. The district commissioner took his chance and, unaccountably, the Lotlhekane settlement and its assets (most of which had, by then, been bought or developed by the Bushmen) were handed over to the Bantu-speaking people of Mopipi and Gomo in June 1940. In the two years he spent on this project, Joyce had also been busy elsewhere in the Bamangwato (now the Central District) and, with Molaba's help, had taught other Bushmen the rudiments of agriculture. It is impossible to assess the impact of their efforts; at the very least it must have improved the morale of the Bushmen and given them some notion of something other than serfdom.

In the wartime and postwar years of shortage of staff and funds, nothing was done specifically for Bushmen. But government officials had become more clearly aware of the special needs of Bushmen and looked after the latter's interests as well as they were able. A decade after the war's end, the government took the view that, with the rest of the country progressing toward independence, a study should be made of the whole Bushman population in order to assess their state and decide the best manner in which their interests could be served and safeguarded in an independent Botswana of the future. Another row broke out about the treatment of Bushmen in the Serowe area, and the district commissioner of Ghanzi, E. H. Midgley, was agitating strongly for effective measures that would assure Bushmen in the western part of the country both economic status and the freedom to decide their own cultural destiny. His divisional commissioner, John Millard, stressed the need for early action in the "gold rush" atmosphere of Ghanzi, which was filling up with land seekers responding to rumors that the ranching scheme was about to be enlarged and more land released for settlement. The resident commissioner, M. O. Wray, CMG, OBE, decided that the Bushman problem was essentially political and should be investigated by an administrative officer who had been trained in social anthropology. At this time I was district commissioner of Ngamiland. During my cadetship in the Colonial Service, I had been lucky enough to spend a good deal of time on work that took me out of my office and into the bush and had often expressed my preference for that aspect of my job. When Wray visited my district in 1956 he asked if I would be interested in doing a survey of the Bushmen. At my request I was sent to the University of the Witwatersrand, where I completed honors courses in linguistics and social anthropology. My terms of reference were decided upon and I began the Bushman Survey in September 1958.

The Bechuanaland Protectorate was a poor country and government resources were slender. No proper financial provision could be made for the survey until the following April, when funds were allocated in Colonial Development and Welfare Scheme No. D. 4721. To tide me over, I was given a scrapped police Landrover and a scrapped caboose (apparently peculiar to the Protectorate, this vehicle, a 5-ton truck, was a sort of heavy-duty camper-van – a small cabin with bunks and cupboards was located behind the driver's cab and the rear portion was a large tray with steel-mesh sides in which fuel, water, and heavy equipment were carried). The government workshops managed to get these two vehicles in running order but were unable to repair the ravages of time and the rough wear that is suffered by all vehicles in that country. I recruited a driver and his assistant (termed, in those days, a Lorry Boy, which epithet was later given cosmetic and bureaucratic overhaul to become Lorry Labourer, Grade V). Experience in Ngamiland had taught me that a driver is a very important member of any expedition. If he could spend a day in a hot, roaring cab, having his arms nearly wrenched from their sockets by a wilful steering wheel protesting against cross-country travel, and then spend the evening on repairs but still remain cheerful, then the journey would go well. Otherwise all was misery and mishap. There was a small, select band of skilled long-distance cross-country drivers, but nobody was going to spare me one of these. They were legendary men who could get through the worst country under terrible conditions in vehicles held together by barbed wire, mopane wood, and black magic. Understandably, some were inclined to be tempermental from time to time. The regular VIP driver, a police corporal whose rejection of conventional discipline drove his commanding officer to strong language and drink, earned undying fame by leaving various resident commissioners, high commissioners, and even higher beings (if such exist) to wearily trudge through miles of heavy sand in blazing heat when he refused to stop for them after getting his truck unbogged from the sand. (In fact, he had no option, for to stop would have been to get bogged again. But it makes a lovely story.) I sought the advice of John Lobatse, doyen of drivers, who had often saved me from my own foolishness when I traveled in his company. I needed a driver. It turned out that he had a kinsman who ran a small fleet of trucks between Tlokweng village and Molepolole. This man was now old and was selling his business. In the last few years his son, Boy Magetse, had helped him. Now, with his father's retirement, Boy Magetse needed a job. Boy had completed part of a mechanic's apprenticeship before going to work for his father. Perhaps he would do. I inquired. His former supervisor in the workshops spoke well of him, and when Boy and I were introduced, we got on well. We went out on a test drive and he showed a certain amount of skill and, joy of joys, a respect for the vehicle. John offered to give him some pointers and I was happy to take Boy on. We went to see his father, and I explained the nature of the job to both of them. Boy would be away from home for a year or more at a time. We would be out in the bush, far from anywhere. There were no fringe benefits, only the meager government pay, and the hours were 24 in each day, and the days seven in each week. We would be hot and hungry, cold and thirsty, alone among the lions and leopards, the hyenas and Bushmen. If Boy fell ill he might die, for there was no doctor. They should think about it before making a decision. John Lobatse could tell them far more than I could; they should ask him all about a bush driver's job. The next morning Boy came and told me he would take the job; his father had agreed and Boy's wife, Miriam, would stay behind and mind the children. We agreed that if he wanted to leave he could do so at any time we were in Ghanzi and that if I wanted him to leave, I would give him a month's notice. As a driver's mate works under his direction, I thought it best if Boy selected his own assistant; John Lobatse and I would veto the choice if necessary. Boy brought his age mate, Phuthego Matsetse, with whom he had been through *bogwera*, the initiation school. That was about all they seemed to have in common. When I began to explain the job to Phuthego, he cut me short. He was an orphan, had no work, no schooling, and no prospects. If the job was good enough for Boy, it would do for him. Boy was then in his late twenties and slightly built. Although highly strung, he had a retiring disposition and was very serious about everything and unfailingly courteous to everybody. Phuthego, big and bland, regarded everybody, from Her Majesty's commissioner to the smallest goatherd, with the same faintly disdainful amusement. In those days his humor ran to quips of near-lethal acidity and practical jokes, each a saboteur's masterpiece. Boy played the guitar well and composed many of his own tunes; Phuthego made up lyrics to many of these. Events during the day provided themes for some, and at the fire before retiring a duet by them made some small thing memorable or reduced to common farce what had seemed like world-enveloping tragedy. Phuthego's lyrics were also a fairly humane way of letting me know when my behavior was becoming a bit too much to bear. Although they were away from their families almost continually and received slender rewards for work that was always hard and sometimes hazardous, Boy and Phuthego stayed with the survey until its official completion at the end of 1965. (I was able to continue unofficially for another year.) They tolerated my shortcomings with great gentleness and, by their support, encouragement, and companionship, carried me through some difficult times.

The two tired, treacherous vehicles with which we started frightened me out of any serious attempt at early exploration of the remoter areas of the Kalahari. They gobbled gasoline and oil and their radiators erupted with the regularity and violence of geysers. At one stage we were so short of water that we had to try and make do on about 5 liters each per day; we found it impossible to do any worthwhile work and did not even feel like eating. Breakdowns of greater or lesser seriousness were daily nuisances or disasters. But it was good training. We learned to work together, to trust one another, and to always be very cautious in our dealings with the desert. Once, 120 km out, the caboose got bogged and fell on a man. He should have been killed. Instead, he was pressed into the soft mud and badly bruised. We learned that the sky would always fall on our heads but that we would find a way of crawling out from under it until the next time it fell.

I had selected Ghanzi as my base because of its central position among the major Bushman areas and its relative proximity to the central Kalahari. This last area had been crossed by the Clifford and Marshall expeditions in 1928 (Makin, 1929) and 1955 (Marshall-Thomas, 1959), respectively. Both had reported the presence of Bushmen who appeared not to have been subjected to any great measure of acculturation. I had previously covered much of western and southern Ngamiland (Silberbauer, 1956) and had visited many of the Bushman communities there. It seemed to me that the central Kalahari offered the best prospects for finding Bushman communities with minimal alien contact. I could not, of course, assume a uniformity of culture among the Bushmen of Botswana; I had seen enough of them to realize that this was not the case. However, it seemed probable that these communities in the central desert might provide an example of aboriginal Kalahari Bushman culture and an impression of the life-style followed by all Kalahari Bushmen before their contact with Bantu and Europeans. This, I reasoned, would give me some insight into the processes of post-contact change that they had undergone.

The language problem was formidable. It was apparent from what I had already seen of Bushmen and from Bleek's (1928) work that radical differences existed among the languages of the Ghanzi and other districts. I had some slight familiarity with Nama, which was sufficient to indicate the complexity of the Central Group of languages and I did not feel capable of dealing with more than one of these languages at a time. I therefore looked for the language that covered the widest geographical area and range of acculturation among its speakers or the language that possessed the greatest measure of central tendency in its structural and lexical features. Although the latter would not serve as a lingua franca, knowledge of it would simplify the task of learning dialects closely related to it.

Accordingly, the first five months of fieldwork were spent on a linguistic survey covering the Ghanzi district west of the Lobatse road, the northern Kgalagari, and southern Ngamiland. At the same time I began a study of conditions on the Ghanzi ranches and Bantu cattle posts on which Bushman serfs were to be found. The ranch study was complementary to the language survey, as the ranching area straddles the boundary between the Central and the Northern language groups (or, in Westphal's 1963 classification, Tshu-Khwe and Bush A). Furthermore, some of the older European settlers, notably Martinus Drotsky, had spent many years with Bushmen, were fluent speakers of the local language, and had extensive, if unsystematized, knowledge of the culture and history of their local areas. Drotsky, in particular, was anxious to teach me, and although his health and memory were deteriorating, he gave most generously of his hospitality and precious knowledge.

From Bushmen on the easternmost ranches I heard rumors of bands permanently living in the central desert and I also encountered part-time squatters who had come to the ranches in the early summer of 1958 in search of water. It became clear that G/wi would be the most suitable language to study. First, because it was understood over a wide area and had the requisite central tendency in many of its features and, second, because there appeared to be a continuum of acculturation among its speakers. Among these there were third-generation ranch laborers, newcomers to ranch life, part-time squatters, and, reportedly, the permanent desert dwellers who were to provide the aboriginal perspective.

I found a G/wi prisoner, /anu, in the Ghanzi jail and recruited the services of Magwe Thamai, a son of one of the jail guards. Magwe was a schoolboy who spoke not only his native Nharo but also fluent G/wi, excellent Tswana, good Afrikaans, and also had a fair command of English. Using Magwe as interpreter, I explained my needs to /anu and started to learn G/wi. We went well enough on items of vocabulary, but even the simplest grammatical structures caused difficulties. My initial approach was to record and to try to analyze texts (i.e., samples of narrative, conversation, etc.) and to derive structures from the analysis. I failed in this because I did not realize the extent of permitted variation in style, which allowed omission of affixes, tenses, and whole words. Reversing the procedure, I tried tailoring the speech samples to paradigmatic utterances. This went better, but it was always hard and slow work for both of us. However, in time I became able to falter along in the language well enough for even strangers to get the gist of what I was trying to say.

With new four-wheel drive vehicles and my smattering of G/wi, we set out into the central desert. Reports of the Clifford and Marshall expeditions had indicated Kgaotwe Pan as a promising area so, for lack of better information, I headed there first. The maps of the central Kalahari were incomplete and inaccurate, being based only on the suppositions of cartographers and the data gathered along the track of the Clifford expedition. (I later found that the position given for Kgaotwe Pan was 60 km out.) I navigated with great care, keeping courses as straight as was practicable in cross-country travel, checked my midday latitudes by observing the sun's meridian passage, and obtained three-star fixes whenever the nights were clear. (For this purpose I used a Kelvin and Hughes Mk. IX BM bubble sextant, the current *Air Almanac*, and Vol. H. of A. P. 1618 *Sight Reduction Tables*, which gave results accurate to within one nautical mile.) Over the years I constructed a map of the central Kalahari (see Silberbauer, 1965), which stood me in good stead in the later stages of the survey when I used a light aircraft for game, vegetation, and other ecological investigation.

Introduction

We set out for Kgaotwe Pan in June 1959. We traveled nearly 3000 km, systematically quartering the area in search of Bushmen. Eventually, close to the limits of our endurance and belief in the existence of Bushmen in the central desert, we came across two men. They had not seen Europeans or motor vehicles before and, terrified, had hidden themselves in a blackthorn thicket. We approached them on foot, stopped a little distance from the blackthorn and offered them tobacco. I tried out my G/wi, inviting them to help themselves, but, like Abraham's ram, they were caught in the thicket. The delicate absurdity of their position reduced them and us to ribald laughter and broke the ice most thoroughly. When we had extricated them, we conversed. After all the trouble of finding them, I did not want to lose them, but neither did I wish to appear to be forcing our presence on them. I explained that I wanted to learn the language, a need that was obvious enough. I also said that I wanted to learn the way in which the G/wi people lived, but this was rather bewildering to them. Then I told them that we had to travel farther to the south but would return in four days. If they were agreeable to our staying with them, we would meet them on our return at this place. Four days later, finding only a silent and empty desert, I cursed my cautiousness. On the off chance that they might still be in the vicinity, we lit a fire to brew some tea and hoped that the sight of the smoke would remind them of our appointment. To my immense relief, they appeared after a while, running across the veld and shouting at us not to go away.

The two men were from an isolated household, the other members of which were a girl (daughter of the elder man, ≠xwa:, and wife of the younger man) and an old woman (≠xwa:'s mother-in-law). ≠xwa: had killed a gemsbok the previous day and the four were temporarily camped at the carcass. Their permanent camp was a few kilometers away. We stayed there for the next three months, three weeks at a time, with a five- to seven-day return trip to Ghanzi to reprovision in between. I made it clear from the start that I would not share our water with the Bushmen. To have done so would have reduced the survey to a water-carrying operation and might have made it difficult for the Bushmen to re-accustom themselves to going without water when we were no longer there. I always felt unhappy about the disparity in our circumstances, but the Bushmen did not seem to resent it.

I always made my camp some distance from that of the G/wi as I did not wish to make our presence any more onerous than need be and because I also found that I needed privacy to nurse my culture shock and to control my frustration when anthropological inspiration ran at low ebb. Most of the G/wi are friendly and sensitive individuals and they became perplexed and disturbed with I was down in the dumps over something. Later, when they got to know me, they realized that this was part of my makeup and attributed it to the ways of the white man. It was, then, useful to be able to sit alone in my camp and get over my depression without upsetting anybody. They visited my camp as freely as I did theirs, but if I sat buried in a book or busy with writing, they spoke to Boy and Phuthego, who were very popular. Their role was difficult. One could not remain neutral and withdrawn from band society, for this was seen as a rejection and, therefore, hostile. At the same time Boy and Phuthego could not become too closely involved lest this too greatly affect behavior and distort what I was studying. They contrived a very satisfactory balance, friendly to everybody but never involving themselves in any manipulative role.

While with the isolated household, I gave out a double handful of tobacco each week to each of the four G/wi, explaining that I had ample and was returning the favor of their "teaching me the language." To ≠xwa:, I gave a share of any meat that I shot. This was a rare event in those days because my rifle was not efficient and I was unfamiliar with the habits of desert game.

I later discovered that people thought ≠xwa: stupid. He was not a good informant to interrogate, for he could not explain things very coherently. But he was a good practical teacher and his enthusiasm for showing me how things were done never seemed to flag. At this time, too, I had no interpreter and made only slow and difficult progress with the language. I began, however, to understand the pattern of life in an isolated household. Perhaps the most useful result of those three months was that the four G/wi came to know and trust us.

The pattern of daily activity changed as summer succeeded winter. At first I spent most of each day with ≠xwa: and Khwakhwa, his son-in-law, interrogating them when they were in camp or following them in their trapping, sporadic hunting (they had only a small supply of rather stale arrow poison left from the previous summer), and digging for roots and tubers. When it became warmer, and activity in the middle of the day was beyond their strength, I left them alone to sleep or rest in the crushing misery of the midday heat, the endless flies, and the blustering wind. We worked together in the mornings and afternoons and talked late into the night around the fire. On one trip we were troubled by lions. I had met lions before in Ngamiland and had been impressed. This first encounter in the central Kalahari was most dramatic. I wanted a good fix on our position and had gone a little distance from the camp to avoid the glare of the fire. When I switched on my torch to read the sextant, it illuminated the face of a male lion who seemed to fill the whole horizon, if not my whole world. Not 2 meters away, he growled very softly. He was so close that I thought for a moment of kicking him in the face. Rationality had deserted me completely; I was stopped from this gross maneuver by a fear of dropping my precious sextant in the sand. Instead, I kept the torch on him and backed off toward the camp, so gently that I doubt if I even bent a single dry leaf. Lord Lion stayed where he was, crouching, occasionally flicking the end of his tail and growling through his half-open mouth. We built up a volcano of a fire and I sat with my dreadful old rifle on my knees, mentally calculating the astronomical odds against being able to do anything useful with it if an inescapable, pressing do-and/or-die need arose. More lions appeared in the firelight and walked around and through the camp until dawn. For the next seventeen nights they kept it, and us, up. None of us was good for a stroke of work in the daytime, other than fetching yet another mountain of firewood. Language work, anthropology, research, everything went by the board, and life squeezed itself into a hard little ball of worrying about the next night. Nobody was physically molested, but psychologically each of us was the mouse that the cat had been playing with. Neither the Bushmen nor any of us could determine what had attracted the lions – 38 altogether and 12 the largest number present on any night. Perhaps they were simply inquisitive.

Later in the summer, just before the rains broke, the four G/wi moved their camp to a spot east of ≠xade Pan. When I returned from Ghanzi they had been joined by three other households. My rapport with ≠xwa: and the others stood me in good stead, and it was evident that he and Khwakhwa had commended us to their colleagues in the band. More households arrived at the joint camp and by Christmas the whole band was assembled. At the end of that summer I recruited Dabe Deneka as interpreter. An old Nharo (immortalized, almost unrecognizably, in Laurens van der Post's *Heart of the Hunter*), he spoke good G/wi, fair Tswana, and an amazing, Rabelaisian version of Afrikaans. With his help, and with a wider choice of informants available in the synoecized band, matters improved and I was able to ask many more questions and probe to greater depths than previously. Interrogation was, however, always difficult. Informants became bored or confused after about 30 minutes, so I kept the sessions brief. The best flow of information was around the fire at night. Several families would gather at one fire and speak of their day and what they would do the next day. Questions about any matter that came up in these conversations were answered eagerly and fully, but I could not direct the conversation itself. Most verbal information came to me in this way, piece by piece. Later, when I came to understand more of their ways, I could fit the pieces together and lead the conversation to some extent by referring to other topics that were logically related to what was currently being discussed. For the rest, information came from watching what people were doing and questioning them a little at a time. I also conducted fairly formal interviews about matters for which I required coherent detail, for example, aspects of the kinship system or the language, and for which I needed to refer to my notes and ascertain that my construction of the pattern accorded with theirs.

I gave everybody in the united band a present of tobacco when I left for Ghanzi and when I returned and pipefuls to informants while we were talking or working together. I bought a better rifle and shot an antelope every fifth to seventh day to keep my men and myself in meat. In addition to Dabe, I recruited a manservant. He saw to my food, a task I was most heartily glad to be relieved of, and helped Boy and Phuthego. I compensated the hunters for their time by shooting a large antelope and sharing the meat with them, via ǂxwa:, who was the nominal hunter of what I shot. I made this arrangement in order to minimize my influence on relationships among the G/wi and because I did not know, initially, what the customary method of dividing a kill might be. My contributions produced a series of exchanges that, as far as I could see, fitted in with the normal patterns and did not have the effect of placing ǂxwa: above his colleagues nor of giving him any particular advantage over them because they recognized his vicarious role.

From April 1960 I began visiting other bands. I usually took a couple of ǂxade men with me as guides and for their good offices in meeting new people. They selected the bands we visited. The encounters were all friendly, even when I had no guides, and all but one of the new bands had already heard of us. Much of that year was taken up in exploring and mapping the central desert and in studying the new bands. I returned to ǂxade band every eight or ten weeks. In 1961 I was appointed district commissioner of Ghanzi and could give only a little more than half my time to the Bushman Survey. I continued to visit other bands and made a trip to the ǂxade people at intervals of ten to twelve weeks. I was overseas for much of 1962 and reverted to the pattern of 1960 on my return. I was then without an interpreter as Dabe had contracted tuberculosis while I was away. He died in July 1964. Another change allowed me to use a light aircraft, which greatly improved the accuracy and extent of observation of the numbers and movements of game animals and vegetation patterns. We constructed an airstrip at ≠xade Pan, which is centrally situated, and, weather permitting, flew sorties for up to seven hours a day for two or three weeks at a time while I had the hire of the aircraft.

The daily pattern of bands assembled at joint camps varied greatly. Women's activities had a fairly regular pattern of foraging, bringing home the gathered food, and cooking it. In autumn they had more time for games and other recreation. The men's day was more varied; some hunted and others occupied themselves with "make and mend" jobs in camp or, in autumn, also with entertainments. I alternated between observing what the men were doing and talking to them about it, joining them on hunts, and filming, photographing, or tape-recording what was going on.

I kept a series of field notebooks and also a journal – the former for recording the piecemeal flow of information and the latter for putting together the pattern of what had happened during the day, as well as for recording routine weather data, radio communication, and lists of what I was to see to when next in Ghanzi.

I did not attend women's activities unless there were also several G/wi men present. When my wife joined us in 1964 she participated in and observed women's activities and most of my information on these comes from her.

When with a band, we managed five or six days' work per week. Hunting for the pot usually took up a whole day as I hunted away from the areas in which the G/wi hunted and consequently spent much time in traveling. My lack of knowledge of game movements in the country that I entered also consumed time. In addition, some time to attend to vehicles and equipment was required.

It is difficult to assess accurately the nature and extent of the disturbance experienced by the G/wi because of our presence among them, particularly the ≠xade people, with whom we spent so much more time. My direct interference with people's activities was maximal when I detained interviewees from their normal business. I did this only after giving them at least a day's notice and after discussing it with them. Even so, it often happened that something would crop up unexpectedly and change our plans, so the interview would go by the board. Filming, photographing, tape-recording, and the presence of vehicles and, later, aircraft were soon accepted as commonplace, although there was abiding interest in the last two. Boys made themselves toy lorries carved out of *Coccinia* roots and men danced a game in imitation of the takeoff and landing of an aircraft. (One dancer became so sophisticated in his version of the game that not only was it clear which type of aircraft he was imitating but, often, which pilot's style of approach was represented.)

I carried a stock of medicines with me and used them to treat the minor and simple complaints of anyone who requested treatment. Doubtful cases were dealt with by radio consultation with government medical officers and, when necessary, patients were taken to Ghanzi and even to Maun. After the airstrip was built, cases could be treated by a doctor flown in for the purpose or patients evacuated to the hospital by air. My wife, who is a qualified nurse, conducted regular clinics for those who requested treatment. The Bushmen reacted to European medicine and treatment with the confident expectation that beneficial results would be produced and considered it as complementary to their own medicine.

A number of Europeans accompanied me on different occasions. With few exceptions, the G/wi enjoyed the novelty of new, vouched-for acquaintances. The exceptions were visitors who were unhappy out in the desert; the G/wi resented their being out of sympathy with the country and regarded the visitors' disquiet as having been occasioned by themselves. They could not understand how anybody could be afraid to be in their country. While I was on overseas leave, a drilling crew sank a series of boreholes, two of which were in ≠xade territory. Boy Magetse accompanied the drillers and did very well in maintaining harmony between them and the Bushmen and in preventing any unhappy developments. A bore northwest of ≠xade was successful and yielded potable water. It was equipped with a pump and engine. At one stage of the drilling operations some 15 drums, with a capacity of 200 liters each, were kept full of water for use by the Bushmen. The constant water supply attracted people from as far away as 160 km, and for a time there was much friction and illness as well as a serious shortage of food, which developed because of the presence of so many people in one territory. The practice of filling the drums was discontinued before I returned, but I was able to see how serious the effects of overcrowding had been.

As far as the central Kalahari part of the Bushman Survey was concerned, my intention had been to divide my time in that region about equally among five or six bands, using them as a sample of the sociocultural system of the desert-dwelling G/wi. I abandoned this strategy when I realized how much additional time would have to be spent in traveling and how difficult it was to achieve and maintain the type of rapport needed to make a satisfactory study of these people. In my experience they were willing informants but, because of the difficulties I had with the language, could not always explain matters very clearly. Their isolation and lack of experience of people unfamiliar with their culture also, I believe, made it difficult for them to put themselves in my place and to understand what I was doing. They could not readily accept that an adult man could really be as ignorant as I appeared to be (and was, of course). Accordingly, I altered my strategy to concentrate on the ≠xade band. This was not only the first band I had encountered but also one of the largest and among those with the least alien contact. As I made contact with other bands and established good relations with their members, I used the information received from them to control the ≠xade data. At the same time the deepening insight that longer contact with the ≠xade people continued to give me enabled me to work faster and with greater precision among the other bands. I do not believe I would have reached the same level of understanding had I persisted with my original intention. I would not have had the opportunity to form friendships or to gain the knowledge of individual personalities that association with the ≠xade people allowed me.

The closer strategy was to follow the daily patterns of activity, influencing these as little as possible. Although this passive role denied me directional control of the flow of information, I believe the picture that eventually emerged was more accurate. This approach did, however, make the work go terribly slowly at times. The G/wi are neither aggressive nor self-assertive toward aliens; consequently, a danger existed that my wishes would be accommodated to the extent that I would almost end up designing their activity patterns for them if I intervened too enthusiastically.

I periodically reviewed progress and, as time passed, was able to discern a pattern emerging among the gaps. Later still, I felt I could risk a small measure of manipulation in order to cover some of the gaps and it was then that I resorted to a greater number of formal, directional interviews or asked people to show me how they performed specific tasks or to demonstrate particular techniques.

I always regarded any information as doubtfully accurate. I acquired the habit when working with ≠xwa: and could not be certain that we had understood one another at all on some occasions. The capacity for misunderstanding diminished with time but never entirely disappeared. (For instance, I only discovered at a fairly late stage that the G/wi words for "hut," "shelter," and "vagina" were

28

distinguished only by tone and that I had been using the wrong one. Although my meaning must have been clear, generally, from the context, the mistake led to some confusion and a good deal of hilarity. This illustrates the dangers facing a single male anthropologist, for the latter word only came up after my wife had joined me in the desert. Her investigation of G/wi obstetrics had revealed one of the finer points of G/wi language.)

What follows in the next chapters is, of course, an incomplete account of G/wi life. Some material was omitted because it is not relevant to the direction of the analysis. More important, the account is incomplete because my knowledge is incomplete. In the years of patient teaching by my G/wi mentors I learned less than they wished me to but more than I had hoped to learn. What first confronted me looked like chaos and I despaired of ever making sense of it. What should, according to anthropological theory, have been a structured organization of society appeared to be nothing more than a series of extemporary arrangements. The harder I searched for the underlying logic of these arrangements, the more offhand and random they seemed to be.

In the early days there was a certain amount of politicking going on in the wings; agribusiness speculators were becoming interested in the central Kalahari and fancied it to be a bovine El Dorado; my investigations were seen as a threat by some very articulate Ghanzi ranchers and by some Tswana and Kgalagari cattlemen. In addition, there were some United Kingdom academics who sought to corner the Bushman market. They all exerted pressure on the Commonwealth Relations Office and on Martin Wray to terminate the Bushman Survey. I was not making worthwhile progress and saw the whole thing folding up in failure, with the Bushmen left to continue as a despised, disadvantaged, and exploited element of the population. At that time it was an act of blind faith to resist the pressures, and I am grateful for it.

In my despair I abandoned theory and tried to learn what my informants wanted to teach me. Now even my most loyal friends expressed alarm that I was wasting my time on nature study. Having started off in forestry, I always had an interest in that direction and what I was doing looked like whimsical self-indulgence. My instruction, or the development of my understanding of it, is reflected approximately in the order of the chapters that follow. I found that I could discern the structure of G/wi society only in terms of its relevance to ecological factors and, therefore, only after I had come to some understanding of the latter. In short, I had started off at the wrong end, or at what was, for me, the wrong end. Others may have found another, better way.

This analysis of the G/wi socioecosystem proceeds from a number of assumptions, namely that we inhabit an ordered universe in which phenomena can be accounted for in terms of other phenomena; that nothing moves except in response to a force and that changes in a system can be accounted for in terms of the flow of energy and information into, within, or from the system – this is a corollary, of course, of Newton's first law of motion ("every body continues in its state of rest or uniform motion in a straight line unless impressed forces are acting upon it"). The third assumption is that it is a human characteristic for an individual, having taken all known and relevant factors into account, to seek to achieve the maximum result from the minimum expenditure of energy. Fourth, I have assumed that it is a human characteristic for members of social groups, having taken into account all known and relevant factors, to seek to prolong the lives of other members of those groups.

The aim of this analysis is not to establish a linear order of causation in explanation of the G/wi socioecosystem. Nor is the analysis necessarily predictive in the sense of making it possible to foretell what development will occur next. The aim is, rather, to explain the system in such a way as to permit a qualitative assessment of the probability of a force being exerted from within the system, the direction and magnitude of the impulse, and the likely response of components in other parts of the system to that impulse. The analysis consists of a study of not only the structures composing the system but also the capabilities of the particular forms of organization, which the structures represent, to respond to, and deal with, the perturbations occurring in the system.

Such a view entails an assumption of coherent articulation within the ecosystem. Ecological theory has developed data-gathering techniques (e.g., Kershaw, 1964) and models to order the gathered data (see Odum, 1971) of the biotic variables in a body of theory.

Social theory has similar means of relating social data. The two disciplines do take each other's data into account but not in the context of a coherent articulation of the two systems about which the disciplines are developed. To illustrate this point: In an ecological analysis, society is represented by inputs of effluent into the ecosystem, without account being taken of the social factors involved in the production and location of the effluent. In a social analysis the disposal of factory wastes is explained in terms of economic and technological factors without taking into account the effects of the pollutants on the ecosystem. The fact that pollutants *do* bring about changes in the ecosystem, and that these changes may be seen as socially undesirable, demonstrates the interrelationship of sets of factors that, for cultural-historical reasons, have been conceptually separated and in the development of the two disciplines have remained heuristically discrete. There is, however, a marked parallel tendency in the development of the two disciplines. As Frake (1962) has pointed out: "If the social system be envisioned as a network of relationships among persons of a social community, then the ecological system is a network of relationships between man, the other organisms in his biotic community, and the constituents of his physical environment. In both cases, the network is woven of cultural threads, and the two networks are, of course, interconnected at many points."

The problem in articulating the two systems is to locate and identify the interrelationships – the points at which the two networks are interconnected, the components of one system that are common to, or interact with, the other. Another problem is to find a denominator that is common to the whole socioecosystem. A perspective that enables one to view the two systems in articulated justaposition is that of general systems theory. One common denominator is energy. Leslie White (1943) formulated a model of energy-flow in the sociocultural system and this has since been refined by successive workers, notably Howard Odum (see H. T. Odum, 1971). Margalef (1968) proposed information as a denominator common to components of the ecosystem and applied this concept to ecosystems analysis. A substantial body of cybernetic theory exists today and application of this to socioecological models should produce considerable refinement. In this study there is limited scope for such application because the gathered data are largely qualitative, not quantitative.

The method adopted in this analysis is to consider the behavior of members of G/wi bands and the products of such behavior as constituting a system – the sociocultural system. The G/wi habitat is viewed as an ecosystem of which the sociocultural system is a component that requires particular emphasis. This emphasis is reflected in the view of the whole as constituting a socioecosystem susceptible to analysis in which energy and information in their various forms and transformations are factors common to all components of the system.

2

The Habitat

The Kalahari

The Kalahari is a basin in the southern African plateau, lying between the peripheral highlands of Cape Province, the Orange Free State, Transvaal, and Zimbabwe, which rise from its southern and eastern margins, and the South West African highlands to the west. It extends, northward from south of the Orange River some 2000 km to the South Equatorial watershed and its greatest width is about 1200 km (Wellington, 1955:52). Some 700 to 1000 meters lower than the flanking highlands, the Kalahari is a vast plain of red, gray, and white fine-grained sands – wind-borne from surrounding regions, the detritus from past ages of weathering (King, 1963:241–244). In a few places rocky promontories protrude through the sand mantle of these Pleistocene and Tertiary deposits. The geology of the central Kalahari has been reviewed by Boocock and van Straten (1962), and it is apparent that this sand-filled basin is an ancient and relatively stable physiographical feature of southern Africa.

Except for the Okavango–Boteti system of swamps and rivers and the Linyati–Zambesi river system in the northwest, the Kalahari is devoid of permanent surface water. Numerous shallow pans and interrupted, ancient drainages concentrate and carry runoff for brief periods during the wet season, and on the eastern edge of the Kalahari seasonal rivers, such as the Nata and the many eastward-trending tributaries of the Limpopo, carry appreciable flows after good rains. Boocock and van Straten (1962) and, more recently, Hyde (1971) have also reviewed the hydrogeology of the central Kalahari and have concluded that potable groundwater is only to be found at great depth in most of the region and that under present climatic conditions the rainfall is insufficient to recharge aquifers that lie beneath more than 7 meters of loose Kalahari sand. "Recharge occurs in areas where the sand is thin or where calcrete or bedrock are exposed at the surface. Where fresh [ground] water is encountered in the central Kalahari, it is probably recharged by water moving down the dip of rocks exposed around the fringes of the Kalahari, or from areas of exposed rock in the Kalahari" (Hyde, 1971:86). From my own casual observation, recharge through a thicker mantle than that mentioned by Hyde seems possible when exceptionally heavy and prolonged rainfall occurs, such as that of 1973–74, but reliable data are not available.

The climate over most of the Kalahari in Botswana is arid. The anticyclone that exists over the northeastern portion of South Africa in winter circulates dry air from the northern interior over Botswana, bringing little or no cloud or rain (Cole, 1961:33). Schumann's (1941) work indicated that when the Kalahari low of summer was accompanied by a cyclone moving northeastward along the Cape and Natal coast, the anticyclone weakened. These conditions could permit the inflow of moist air from the Indian Ocean to the interior of southern Africa. Although most moisture is precipitated over the high eastern portion of the plateau, the Limpopo-Sabi depression provides a rift in the escarpment through which moist air can penetrate as far inland as the Kalahari in Botswana with comparatively little reduction of moisture content. The moist east wind may be established from an offshore low just north of the Limpopo-Sabi valley in accordance with Schumann's hypothesis, but a more acceptable explanation of Kalahari summer rain is that the easterlies are established on the northern flank of a new, intensifying high forming in the wake of a low that has moved well beyond the valley. Lows are small in extent, rather weak, and move quite rapidly. The highs are more probable generators of wet-weather conditions in the Kalahari, as they are larger and stronger and, being slower-moving, produce and sustain the appropriate wind direction for the length of time needed for the moist air to travel from the coast into Botswana. Since the advent of weather satellites, it has become apparent that the northeastward moving lows are not as important as had been thought in Schumann's day – they fortuitously entered the area of most intensive observation and were therefore noticed more often than were other lows – but small lows move in all directions away from the coast (Paul Annette, pers. comm., 1973).

Because the inland bodies of surface water are not large enough to replenish the drying air of the easterlies, the Kalahari in Botswana has a mean annual rainfall of from 250 to 350 mm with an annual coefficient of variation ranging from 50 to 80 percent (Pike, 1971b:73). The wet season normally starts when surface temperatures have risen high enough to establish a strong Kalahari low and, in response, a strong convergence mechanism, whose needs can be adequately met only by drawing air from distant parts (P. Annette, pers. comm., 1973). Wellington (1967:28 ff.) has postulated a relaying process in the movement of the summer rains, as a consequence of which the rain

Table 2. *Annual rainfall (in millimeters) at two stations, 250 km apart, west and east of Central Kalahari Game Reserve, 1961–1965*

Station	Mean	1961	1962	1963	1964	1965
Ghanzi	401.4	375.0	246.0	771.5	168.5	294.7
Rakops	471.6	446.0	431.5	436.5	156.7	102.5

season is established later and ends sooner in the west than in the east. In the central Kalahari the wet season is normally only established toward the end of December. In some summers the equatorial air mass reaches into the Kalahari during southward displacement of the Intertropical Convergence Zone (Pike, 1971b:72). This occurred in 1973–74 when the wet season lasted from early October until May and falls of more than five times the annual mean were recorded on Ghanzi ranches. Such summers are freakish and rare events and are followed by dramatic but short-lived changes in flora and fauna. As I have said earlier, it is probable that many of the discrepancies among early travelers' accounts of the state of surface water, the species and numbers of animals encountered, and the condition of vegetation in specific seasons (see Campbell and Child, 1971) may be explained by the occurrence of similar aberrant floods.

The southeast trades on the eastern flank of the South Atlantic high blow offshore and block the passage of moisture-bearing winds into the interior across the western coast of southern Africa. The central Kalahari is, therefore, dependent on the easterlies for most of its rainfall. However, light showers from the west usually presage the onset of the wet season: These probably represent the remains of moist air drawn up from the Cape by the strengthening Kalahari low before the easterly trend up the Limpopo-Sabi gap becomes established.

Rainfall in the central Kalahari is erratic in both its annual and local variation. Pike's isohyetal map (1971a:fig. 6) shows a 350 mm mean annual rainfall with 60 percent seasonal variability. Some indication of local variation is given by comparing the annual falls of two stations some 250 km apart on the western and eastern sides of the Central Kalahari Game Reserve (Table 2).

No systematic data are available for the central Kalahari; *faute de mieux* those of Ghanzi must serve as a basis for appreciation of climatic conditions in the region as a whole. In my experience, a greater temperature range is experienced at ≠xade Pan, where the extreme minimum measured by me was −13 °C and the corresponding maxi-

Table 3. *Average interdiel variability of maximum temperatures at Ghanzi*

Month	Variation (°C)	Month	Variation (°C)
September	1.9	March	1.9
October	1.8	April	1.6
November	2.2	May	1.4
December	1.9	June	1.8
January	1.8	July	1.6
February	1.6	August	1.6

Data from R. J. Andersson, pers. comm., 1966.

mum was 48 °C. My own intermittent observations between the summer of 1958–59 and the end of winter 1966 suggest that the central Kalahari receives less rain than do the more northerly Ghanzi and Rakops. Pike's map (1971a:fig. 6) supports this notion.

Despite the great seasonal variation in temperatures, the short-term temperature pattern is very regular (Table 3).

The frost data for Ghanzi (Table 4) must be regarded as a conservative indication of central Kalahari conditions. The station is on the Ghanzi Ridge, more than 200 meters higher than the central Kalahari and noticeably less subject to frost.

Nearly all rainfall is from violent, localized thunderstorms that occur in the afternoon and sometimes persist as residual storms throughout the night into early morning. There are no available records of the intensity of single storms, but Wellington (1967:36) quotes South-West African 24-hour maxima of up to 130.1 mm and concluded that the intensity of falls in that part of the Kalahari does not closely correlate with annual totals but is irregular and highly variable. In my observation the localized nature of convection storms and the variability of rain intensity are such that areas quite close to one another may receive widely differing total rainfall during the season. During favorable conditions of moist-air inflow, storms occur daily over periods lasting a week or ten days, which are then followed by dry spells of up to five days. A shorter phase of less frequent storms then follows, after which there is a longer dry period. There are usually three or four such episodes during the wet season.

Most of the Kalahari lies within the 80 percent isohel (Wellington, 1955:226, fig. 64) and for most of the year the sun beats down with little or no cloud to give shade from its heat. Summer temperatures are aggravated by the high reflectivity of the bare sand (reflectance

Table 4. *Occurrence of frost at Ghanzi*

Average first date	June 9
Average last date	August 1
Average duration of frost period	53 nights
Extreme first date	April 27
Extreme last date	September 23
Years of data	37
Number of years in which frost occurred	35

Data from R. J. Andersson, pers. comm., 1974.

values vary from 50 to 85 percent with the color of the sand; R. J. Andersson, pers. comm., 1966). Sand temperatures as high as 72 °C have been recorded (Cloudsley-Thompson and Chadwick, 1964:15). Surface temperatures under trees and in the open differ by as much as 30°C (Leistner, 1967:25). Humidity is low for most of the year (Figure 2), which, although reducing the subjective experience of discomfort, causes a rapid evaporation of moisture. Pike (1971a:Table V) reports an annual mean open-water evaporation of 1726.1 mm for Ghanzi. The mean evaporation rate exceeds average rainfall by more than 100 mm in each of the months of the wet season (Pike, 1971a:Table V, and Andersson, pers. comm., 1974). Water-filled pans, or playas, soon dry up after the rain has passed, and many plants transpire at lethal rates, quickly wilting in the dry spells that interrupt the rainy seasons in many years.

The Kalahari is not a desert devoid of vegetation. Its thick sand mantle supports a varied, if sparse, cover of grasses, vines, forbs, shrubs, and trees. These evade drought by either completing their reproductive cycles during and shortly after the wet season or are dormant during the dry season. Almost all perennials are markedly cold/dry deciduous and lose their foliage between autumn and during, or shortly before, the early stages of the wet season. A few families, mainly succulents (e.g., *Aloe* spp. and stapeliads), have a limited capacity to withstand the long dry periods without taking the evasive measures of other plants.

Away from the river systems the differences between one part of the Kalahari and another are small. In the absence of topographical diversity, plant life lends variety to the flat scenery of what has been called the Thirstland, which seems so monotonous when one first observes it. The small changes – one type of thornveld to another, scrub plain to grass plain, and so on – are due principally to differences in soil or, over greater distances, to variation in rainfall.

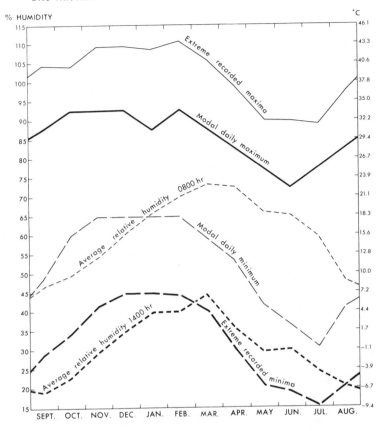

Figure 2. Monthly variation in maximum and minimum temperatures and relative humidity, Ghanzi. (Data supplied by R. J. Andersson)

The Central Kalahari Game Reserve

In response to recommendations I made to the Bechuanaland Protectorate government on April 28, 1960, some 52,000 km² of the Ghanzi district east of the latitude of Great Tsau Hill (Sonop Koppies) was proclaimed a game reserve in 1961 (Figure 3). This step was taken to protect the Bushman inhabitants of the area; in the late fifties illegal hunting by non-Bushmen from outside the area posed a serious threat to the hunters and gatherers who depended on the game herds for part of their livelihood. Hunting by Bushmen was not restricted in any way, but entry by non-Bushmen was controlled and the danger from poachers diminished.

37

Top: A wildebeest killed by drought. Bottom: G/edon!u waterhole drying up toward the end of the wet season.

The reserve lies approximately in the center of the Kalahari. It is low country, north of what Passarge (1904) termed the *Bakalahari Schwelle* (bakalahari rise), east of the Ghanzi Ridge, south of the Mabeleapudi–Dinokwane range of hills, and west of the Ngwato Escarpment. Most of the reserve constitutes a shallow, tilted basin, which drains into the Lake Xau–Boteti system. The main axis of this drainage is along the Okwa and Deception valleys, which appear to be fossil remnants of great rivers that once flowed from the western

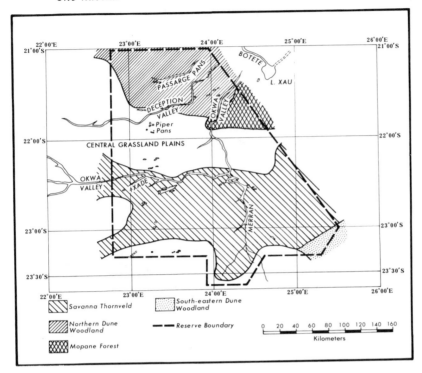

Figure 3. The Central Kalahari Game Reserve.

highlands into the Makgadikgadi Depression via Lake Xau. The reserve, although part of the Kalahari biome, is a distinct ecological
unit, a transitional ecotone linking the woodlands of the north with
the thornveld and scrub plains of the south. It lacks the comparatively rich grasslands of the southern Kalahari, where deep-sunken
pans and shallow groundwater have permitted settlement by Bantu-
speaking pastoralists, who established small cattle posts about the
wells that they dug in the pans. Groundwater also lies shallow along
the Ghanzi Ridge, where European cattle ranchers were settled by
Cecil Rhodes in the nineties (see Silberbauer, 1965:114 ff.; Gillett,
1970:52–55). East of the reserve, along the Ngwato Escarpment, good
grazing and shallow ground water have encouraged settlement by
other Bantu-speaking pastoralists. The reserve, then, was sur-
rounded by black and white settlement but, at the time of which I am
writing, had not itself been occupied by anybody other than Bushmen and a few small groups of seminomadic Bantu-speaking people
who intermittently penetrated the area with their stock for short periods in favorable seasons.

39

The habitat

No systematic study of the reserve's soils has yet been made. Blair Rains and Yalala's (1972) analyses of soil samples and vegetation at 11 sites around the periphery of the reserve are helpful. The soils were all very fine-grained sands (average grain diameter around 200 μ), rather acid (median pH 5.9), and very deficient in potassium. Whiteman characterizes the soils as virtually structureless "sand with very low levels of organic matter, resulting in a low cation exchange capacity and low ability to retain nutrients, and susceptible to windblow" (1971:116). He also mentions their low water-storage capacity – seldom above 25 mm of available water per 300 mm of soil. Blair Rains and Yalala found that in the dry conditions at the end of the growing season plant litter is mostly carried away by wind or eaten by termites. The organic carbon content of the soil samples was on the order of 0.2 percent, which is very low. Although their survey was made during February and March 1969, when the wet season was advanced and the vegetation well grown, plant coverage of sample areas only approached about 30 percent. In other seasons, especially early summer, there is much less plant cover and even more of the ground lies bare.

The Okwa and Deception valleys, the most prominent topographical features of the reserve, divide it into three distinct regions: the dune woodlands, which lie between the northern boundary and Deception Valley; the central grassland plain between the two valleys; and the thornveld south of the Okwa. Each region has distinctive features of topography, vegetation, and fauna.

The dune woodlands

This region appears to have been swampland until very recent geological times. It is dotted with numerous large pans, the drainage axes of which point toward the Boteti, and the long valley of the Passarge Pans is strongly reminiscent of the major flood channels of the Okavango Swamps farther to the north. The floors of these pans are almost level with the upper terrace of the Boteti (see Jeffares, 1932), and in years of exceptional flow, its floodwaters would reach back into these pans were it not for the barrier of sand that now lies between them and the river. Possibly, in the past, vegetation blocked the flood channels of these putative swamps, and the low-lying floodplain then dried out; certainly, the area of free water surface of the Okavango–Botete system was once much greater than it is at present (van Straten, 1963:31). Sand carried by the prevailing northeaster could have accumulated in the desiccated region, forming the dunes that reach up to 20 meters in height and now dominate the local scenery. The dunes have been effectively anchored by vegetation (see Figure 4) and today provide shelter from the strong winds of early summer. The soil of the interdune flats is richer in humus (Blair Rains and Yalala, 1972:67, 92) and nutrients and has better texture than the soil of the country farther south. As the rains enter from the northeast, the wet season here starts a little earlier, lasts longer, is more reliable, and yields a greater total fall than is the case in more southerly regions.

The better soil and rainfall of the woodlands support a more varied vegetation than is found farther south. The dune crests are topped by dense spinneys of sterkbos trees (*Terminalia prunioides*), various shrubs (*Grewia, Bauhinia,* and *Dichrostachys* spp.), and sour grasses. Sweeter grasses grow on the interdunal flats. In places the flats are covered by almost impenetrable brakes of paperbark (*Acacia fleckii*) and clawing blackthorn (*Ac. mellifera*) and the shrub species found on the dunes. Elsewhere the valley floors have a parklike aspect with *Ac. nilotica* and *Ac. giraffae* standing tall and widely spaced above the grass. These and other thorn trees (e.g., *Zizyphus* sp.) also grow in small woods about the occasional steep-sided wallows where water accumulates and stands for some weeks after good rain.

The pans are large, flat expanses of tufaceous soil, hard-baked and bare in the dry season but slippery and treacherous when wet. In the rainy season the softened tufa is easily trodden out by game animals, which favor the pans because the palatable pan kweek grass (*Sporobolus tenellus*) grows there and also because the clear field of view protects them against predators. The animals congregate in the center of a pan, scuffing up the mud in search of roots or carrying it away on their hooves, and eventually excavating small depressions. Rainwater gathers in these depressions, attracting yet more animals and, consequently, more mud is carried away. Eventually the depressions are consolidated into one large waterhole. Seeds carried by animals and on the wind lodge in cracks in the drying mud and, in time, a small shading grove grows up around the pool.

The comparative abundance of sweet grass and the plentiful shrubs with palatable foliage support many grazing and browsing animals. Dense thickets and open pans offer them shade, shelter, and a measure of safety from predators. As climax communities, these areas have greater biomass, offer a wider array of niches, and have a greater diversity of species than other biotic communities elsewhere in the reserve. The populations are somewhat more stable and the behavior of the animals comprising them is more predict-

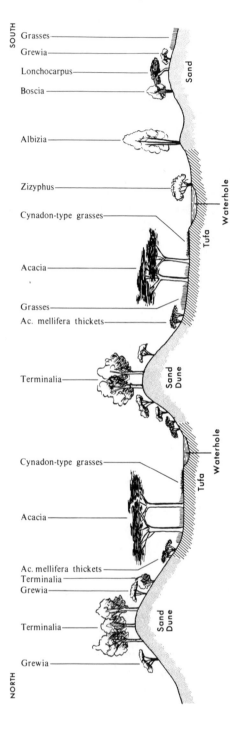

Figure 4. Idealized north-south transect of 8 km of dune woodland. The vertical interval is greatly exaggerated.

able. Unfortunately, a growing number of poachers exploited these characteristics in the late fifties and early sixties and threatened the integrity of the animal and plant communities by shooting and scaring game and by firing the bush to drive out their prey, thus destroying large tracts of woodland each winter.

The northern boundary of the reserve was fenced in the late fifties to prevent the movement of game animals between Ngamiland and Ghanzi district and thus restrict the extent of outbreaks of foot-and-mouth disease. Economically, this was a rational step, as the disease, which was then believed to be carried by all game animals (Henning 1956:895 ff.), affects cattle on which the economy of Botswana was then almost totally dependent. Subsequent research has shown that central Kalahari game herds are normally free of the disease, which, it appears, is only perpetuated in buffalo herds (Falconer, 1971:154–156). The movement of game animals into and out of the reserve is not random but holds to a fairly regular pattern of purposeful migration in search of feed and water. Poverty resulting from overcrowding against the fence caused high mortality among migrating animals (see Silberbauer, 1965:20–21).

The central grassland plain

In their provisional classification, Weare and Yalala (1971) have grouped the grasslands and dunelands together as Northern Kalahari Tree and Bush Savanna. Blair Rains and Yalala (1972) have retained this grouping. Although the designation is probably valid for their purpose of vegetation mapping on a national scale, there are sufficient floristic and other ecological differences to distinguish between the dunelands and the grasslands in this study. I also differ from these authors in that I extend the Mopane Line some 90 km farther to the southwest.

The transition from dune woodlands to grassland is fairly gradual. The woodland species become progressively smaller and less dense southward toward Deception Valley* with larger and larger belts of

---

* So-called for two reasons: First, when traveling along it one is deceived into thinking that every bend is its end and, second, after Deception Pan, which stands on the valley's right bank. This pan is rich in magnesium sulfate, the crystals of which cause an odd refraction of light that gives to the surface of the pan the appearance of water. The illusion persists to within some 25 meters of the rim and is complete, even to the sight of ripples and wind-lanes, splashes around the feet of running antelope, and a damp stain about the pan's periphery.

43

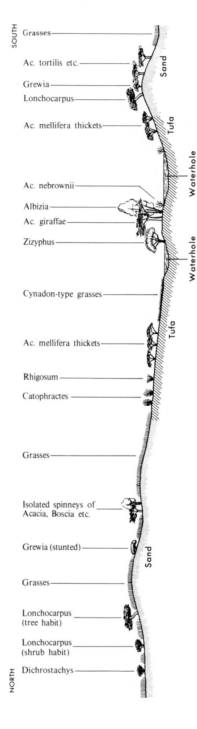

Figure 5. Idealized north-south transect of 6.4 km of scrub plain and pan. The vertical interval is greatly exaggerated.

scrub between them (Figure 5). The dunes also fade and their crest vegetation changes to scrub.

The plain is a very slightly undulating expanse of short, sparse grass and small shrubs. Its monotony is interrupted only by occasional, shallow, narrow pans, a few belts of low woodland, and solitary trees, which seem tall at a distance but, when one approaches them in the hope of finding shade, turn out to be no more than 3 or 4 meters high.

There is some difference in the length of the wet season between the northern and southern parts of the plain. This is not reflected in the vegetation, which is uniform, but in the movements of game animals. In most years the pans in the north have water some weeks earlier than those in the south, and they dry out some weeks later in the season. The Piper Pans complex, which might have been the westernmost outlier of the putative swamp discussed earlier, interrupts the course of Deception Valley, which loses itself for some miles in the maze of pans. The complex covers about 1000 km² and is the focus of movement of huge herds of antelope in the wet season. Grazing and browsing are good in most seasons and the grasses (*Stipagrostis ciliata* and *S. obtusa*) on the pans seem to grow as fast as they are grazed by the herds. The fact that a party of "sportsmen" shot 34 lions on Piper Pan itself in one weekend in the winter preceding the proclamation of the game reserve gives some idea of the great number of antelope that were still there, long after the summer rains. In the preceding summer I drove around a mized herd of gemsbok, eland, and hartebeest that covered an area 8 km long and 5 km wide. At that time the pans also abounded in wildebeest, kudu, springbok, giraffe, lion, leopard, and cheetah. There were scores of black-backed jackal and bat-eared fox and numerous vultures and raptors.

This complex of pans has a few botanical oddities. The spiny shrub *Sesamothamnus lugardii*, which looks like a 3- to 4-meter high baobab, grows on the south side of Piper Pan itself. As far as I could discover, *S. lugardii* only occurs elsewhere a long way to the east and, reportedly, on the Gobabis plateau (Codd, 1951:168). A species of *Hoodia*, which Kew botanists tentatively considered to be a new species (E. Milne-Redhead, pers. comm., 1962), grows on Hoodia Pan, which is also part of the complex.

The pans are more distinctly the focus of life in the grasslands than in the woodlands. Seeds blown or carried into the hollows where water collects grow to form denser and taller thickets than are found on the open plain and offer the only real shade and shelter to the larger animals. The trees provide safer nesting sites than can be found for miles around and the gregarious species of antelope favor the pans because their lawnlike surface lacks the cover given by surrounding scrub to stalking predators. Most important is the fact that the relatively impervious tufa floors of the pans are the only places where rainwater does not immediately drain away but stands for some time before evaporating, providing the only supply of water to be found on the plains.

The grasses are mostly sour types, but nearly all shrubs have palatable foliage and are browsed by antelope, giraffe, and a host of smaller herbivores. With few trees to suppress their growth, vines and bulbs grow more prolifically here than in the woodlands. Many of these plants have leaf patterns that funnel rainwater down to the bulb or tuber beneath. The storage organs, swollen by rainwater, furnish man and animal with a substitute for water during the nine or ten months when the pans are dry. Although the grasslands do not support the density and variety of animal life that occurs in the dunelands, there is enough food and fluid for small herds of antelope to remain on the plains throughout the year.

The plain is on the route taken by game herds when they migrate from the northeast in summer on their way to the southwestern Kalahari. If the rains around the Piper Pans complex have been good, some of the herds are diverted and spend their summer in that area. The others move on, some to penetrate as far as Gemsbok National Park, which lies partly in the Cape Province. Those passing through travel fast across the plain, covering as much as 50 km in a day. The return journey is more leisurely, and the big herds spend up to two months in the central part of the reserve in autumn.

Apart from the migrants, there is a permanent population of all game animals. This is smaller than the migrant population and is more thinly dispersed.

The thornveld

The fossil valley of the Okwa marks the northern boundary of the thornveld, which Weare and Yalala (1971) term Central Kalahari Bush Savanna and which Blair Rains and Yalala (1972) have classified Central Kalahari Transitional Savanna, that is, as being transitional between the southern savanna and the northern type with its taller and denser growth of shrub and greater number of trees.

The present profile of the Okwa valley (see Figure 6) is partially obscured by a mantle of sand some 30 meters thick (indicated by samples from boreholes sunk near ≠xade Pan). Only the nickpoint,

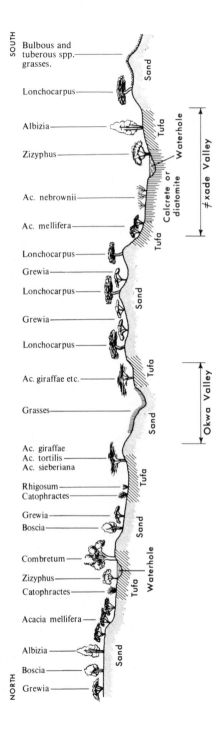

Figure 6. Idealized north-south transect of 24 km of the Okwa-≠xade region. The vertical interval is greatly exaggerated.

averaging some 10 meters deep and up to 500 meters wide, is to be seen clearly. Careful examination of the banks reveals the successive terraces, which stand some 30 meters higher than the channel bed. In its heyday the Okwa appears to have been a river of proportions approaching those of the present-day Zambesi (C. H. Jennings, G. Lamont, pers. comms.). The Okwa has several tributaries; all but one – the Hanahai – enter from the south. These right-bank tributaries have relatively impervious floors of calcrete and diatomaceous tufa. This type of surface is also common along the first and second terraces of the main valley, and good waterholes can be found along the terraces and tributaries during the wet season. Large sinkholes exist in the calcrete and diatomite floors of some tributaries (e.g., ≠xade and G/edon!u valleys). The G/wi say these were used by their forefathers as giraffe traps. It is easy to imagine how, in a well-coordinated drive, giraffe that came to drink at waterholes could have been stampeded into the sinkholes. I could not discover why the G/wi had abandoned what seemed to be a thoroughly practical method of getting giraffe meat to pursue these extremely hardy animals with bow and poisoned arrow, a much more arduous task.

Apart from the pans and valleys and a few large, isolated dunes, the topography of the thornveld is as flat as that of the plains. The vegetation, however, is quite different, possibly because the thornveld sand is richer in nutrients here. Most of the country is covered by tracts of fine-leafed woods in which trees seldom exceed 7.5 meters in height. Apart from the limiting factors of water and soil nutrients, the high winds and lightning of summer storms sooner or later destroy those specimens that are appreciably higher than the surrounding bush. The commonest species are the thorn acacias (*Ac. giraffae, Ac. gillettae, Ac. sieberiana*), smooth round-leafed *Albizia anthelminitica,* and witgat (*Boscia albitrunca*). Trees here do not have the thicket-forming habit of farther north but grow in more evenly spaced, mixed stands. The wag-'n-bietjie thorn (*Zizyphus mucronata*) and the leadwood tree (*Combretum imberbe*) grow in the calcritic soils occurring in and around pans. Appleleaf (*Lonchocarpus* sp.) and sand-yellowwood (*Terminalia sericea*) are widespread shrubs and some specimens reach tree size. Pypsteel (*Ochna pulchra*), wild orange (*Strychnos* sp.), and wild seringa (*Burkea africana*) are common in the eastern parts of this zone. In places they grow widely spaced, giving the appearance of parkland. In the far east the woodlands are supplemented by extensive stands of mopane (*Colophospermum mopane*).

The shrub layer consists of stands of sand-yellowwood, appleleaf, several species of raisin bush (*Grewia* spp.), bride's bush (*Bauhinia macrantha*), and, in tufaceous soils, the spiny *Rhigozum brevispinosum* and pomegranate bush (*Catophractes alexanderii*). The shrubs form extensive stretches of scrub and also grow mixed with trees. A low *Commiphora*, the wild plum (*Ximenia caffra*), and the Kalahari Christmas tree (*Dichrostachys cinerea*) form occasional brakes.

The plants of the ground layer include vines, creepers, tuberous and bulbed species, and a great variety of ephemerals. The grasses tend to specialize according to soil and community types and are either deciduous or ephemeral. The response of all vegetation to rain is rapid, but the response of the forbs is quite dramatic. What was a bare, black and dirty-white waste of seemingly dead trees and sand is mantled with green within a few days after good rain and is soon dotted with the many colors of an astonishing array of flowers.

The mammals of this zone are less plentiful and varied than are those of the dunelands. Migrating herds do not congregate here to the extent observed farther north. The populations are, however, more stable than are those on the plains in that they tend to move only when disturbed or when they have exhausted the local supply of food; consequently, they remain in one vicinity for weeks and even months at a stretch. Among the antelope, wildebeest and springbok are coarse grazers and browsers, and large herds virtually denude the veld before trekking to new pastures. The bare land they leave behind is subsequently avoided by other herbivores. Giraffe are usually found in family groups, although aggregations of up to 48 were encountered. These grazed and browsed within small areas before moving on. Kudu are shy and keep to the thicker bush, where they will tolerate a surprising degree of disturbance before abandoning one wood for another. Once worried, they take a long time to get over the upset and are even more wary than before. They run in small herds and the bulls separate from the cows after the mating season, so the food resources of a run are but slowly exhausted by this species.

Steenbok and duiker are nongregarious, running singly or in pairs. Tenacious of locality, even to the extent of returning after fire has destroyed the pasture, their regular patterns of daily movement are exploited by G/wi hunters, who set snares for them (see Chapter 5).

Hartebeest are perhaps the most mobile of the antelope in this area, favoring the open stretches of grass and low shrub. Eland form herds of up to 20 in number, some of which remain in the same locality for months at a time if not disturbed too much. Gemsbok form larger herds here and tend to be more mobile than eland, often following a circular pattern of local migration in which they return to the same area after some weeks' absence.

Although this southern zone has not the wealth of plant and animal life found in the dune woodlands, it appears to offer a more secure habitat for hunters and gatherers in its greater variety of esculent plants, which are available in all seasons of the year. Bushman informants suggested, and my own observations tend to confirm, that the rich stock of food plants found in the dunelands in some seasons are not always so plentiful.

3

The G/wi Universe

For heuristic convenience I have taken the theology of the G/wi as a starting point in explaining how they conceive of the universe, how they see the world, and how they construe their experience of it. At least as far as the G/wi are concerned, I concur with Clifford Geertz's view that the religious beliefs of a people represent "their most comprehensive ideas of order" (1966:3). G/wi theology is central to their world view (or so it seems to me) in that it articulates their diverse areas of knowledge and belief, their values, and the structure of their logic in a coherent whole.

Religious beliefs

The religious beliefs of the G/wi are not formulated in doctrinaire creed but are part of their general knowledge. Consequently, the same variation exists in the details and extent of theological knowledge as exists in other aspects of their general knowledge. The following is a synthesized consensus account.

A being, N!adima, created the fabric of the universe and the prototypical specimens of all life-forms. He ordained the manner in which each life-form should live, breed, and die, and, within fairly broad limits, he also ordained the relationships that exist both between different life-forms and between life-forms and their environment. N!adima set the sun, moon, and stars on their courses and made and put in motion the cycle of the seasons and the weather systems. The phenomena that constitute the environment and the order that prevails in and among these phenomena are all seen as N!adima's creation and as being subject to his will. An account of the beliefs and attitudes concerning N!adima is, therefore, necessary to an understanding of how the G/wi view their environment.

N!adima is the supreme being in the universe; he is not only the creator and the owner but the only being not subject to the will of another, and all others are subject to his will. As the owner of the universe, he has the freedom to do with it what he will within the

51

limits of the system that he himself imposed. To exemplify: He can and may withhold rain from an area, but he cannot make rain fall unless the normal portents and precursors of rain first appear. The integrity of the systems that he ordained appears to be inviolate. But within the limitations imposed by his own acts, N!adima is omnipotent. He is also omnipresent, eternal, and omniscient.

He is known to be anthropomorphic, or, at least, human shape is one of the forms he assumes. This information is gained from his rare and random appearances in dreams, in which he instructs men and women in the solution of problems and shows them ways of overcoming adversity. However, the anthropomorphic aspect of N!adima does not mean that he is thought to have created man in his own image. His human characteristics are only part of his identity, the totality of which is beyond man's comprehension and beyond comparison with man.

N!adima and his wife, N!adisa (the feminine form of his name), live in the upper region of the three-tiered universe, above the visible sky. N!adisa's offspring include at least all of the mammal species that occur in the central Kalahari and possibly all of the species comprising that fauna. The mode and means of the deity's family life are largely unknown, but there is a belief that they are all vegetarian (i.e., they do not kill to eat) and that an abundance of water and plant foods exist in their country. Informants among the G/wi living on the Ghanzi ranches and some !xō: and Nharo Bushmen in the far west of Ghanzi district gave me more detailed accounts of the daily life of the deity. They expressed belief in an afterlife spent with the deity in his sky country, and their more extensive knowledge appears to be associated with a missionary-derived identification of N!adima with the Christian God and a belief in Heaven. This identification and the prospect of an afterlife in the country of N!adima are not part of the theology of the desert-dwelling G/wi.

N!adima's purpose in creating the universe is unknown and beyond man's understanding. However, whatever it may have been, it is certain that it was not to create an environment for the special and principal benefit of man nor of any other life-form. Although my representation of the G/wi cognitive map stratifies life-forms in concentric rings that are nearer to or farther from man, there is not a matching stratification of authority over life-forms nor of rights to resources in the environment. All N!adima's creatures have equal rights to existence. Each species, with its own characteristics, has a place in N!adima's world. Some prey upon others and some have to struggle harder than others to survive, but none is thought to be uniquely favored by N!adima and to have been set above others by him (exemplified in the belief that the offspring of N!adisa include creatures other than man). All are his creatures, which he suffers to exist for as long as he chooses, leading the style of life that he ordained for them and making use of what he has made available to them.

N!adima is the giver of life. Life is conceived of as a state or condition of an organism and not as an entity in itself. There are two aspects of the process of reproduction: the part played by the parents in growing the fetus, egg, or seed and N!adima's part, which is to vivify it in a separate and mysterious act. Life is also terminated by N!adima, to which end he employs natural phenomena such as disease, animals, fire, drought, and so on. His reasons for taking life are largely unknown. It is believed that he can be angered by behavior disrespectful to him, including belittling or wasting what he has created; but without any orthodox prescription of good behavior, there is no promise of immunity from his displeasure or caprice. For no knowable reason he may "grow tired of seeing a man's face" and end that man's life, even though it may have been blameless. Singular misfortune, even death as a result of an unusual accident, are attributed to N!adima's having intervened in the operation of natural systems and deflected them to serve his purpose in ending the life of the afflicted individual. In the complex web of relationships between N!adima's creatures, the misfortune of one may be the salvation of another. It is understandable that the aspect of misfortune should have been prominent when a lion made a freak attack on a young child. It was, however, a significant contrast of circumstances when a band of armed hunters, ready to set out for their day's hunting, were able to dispatch at once a hartebeest that had blundered into camp. This was seen as N!adima's hand against the hartebeest rather than as a sign that he favored the hunters. It was not that the happenstance was not welcomed and enjoyed but that N!adima's intervention on behalf of the men would have taken a different form. Had N!adima chosen to favor the men, such a sign would have been in the form of a dream appearance in which one of the hunters would have been told where to find the hartebeest.

N!adima is remote from his creatures in the sense that they cannot communicate with him to influence his behavior in their favor. In his omniscience, he knows their wants and their doings; if he wishes to favor them with assistance by granting them knowledge or by deflecting natural systems, he does so of his own accord and for his own reasons. The G/wi accept their largely passive role and have no prayers, hymns, worship, sacrifice, or acts of celebration and praise

53

of N!adima, although they do address the sun, moon, and thunderstorms (see following). These are seen as particularly powerful creations and the G/wi make their addresses to dispel any suspicion in N!adima that they may be disrespectful to these his (presumably) prized creations.

Another being, called G//amama or G//awama, is polymorphic, that is, he is capable of assuming different shapes at will, and is also capable of making himself invisible. The G/wi have no account of his origin, and he has no particular abode, but, as an informant said, "He lives everywhere and nowhere – he moves in whirlwinds and moves in the sky among the stars and sits among people in their camp." G//amama is neither all-powerful nor omniscient. He apparently can take life but cannot bestow it, and his power to manipulate natural phenomena is less than N!adima's. G//amama effuses a generalized evil, which is contained in small, usually invisible slivers of wood, *kxaog/wag//wa* (small arrows), that he showers down on the band from the sky. The slivers lodge in the women and the evil then diffuses through the band. Its effect is shown in the people's increased irritability and in the greater number of misunderstandings and minor misfortunes that arise in the band. The frequency with which G//amama is believed to strew his evil arrows is roughly proportional to the frequency and extent of interaction within the band. The risk is least when the band is in the separation phase and interaction is confined within the isolated household. He is meanminded and cunning and knows that damage is greatest and is felt most keenly when people are together, enjoying the happiness of one another's company. During visiting season in autumn, when the camp may contain thirty or more households and food is so plentiful that people can afford to spend much of their time talking, dancing, and playing games, exorcising dances are performed three or four times a week to counteract the evil that G//amama bestows.

G//amama is innately hostile to man and habitually, but erratically, tries to wreak harm by sending disease or other misfortune in addition to his slivers. Through their own discoveries and with the help of N!adima, the Gwi have developed the use of countermeasures, such as medicines and dances, to frustrate G//amama's efforts. In their dealings with him, the G/wi play a defensive role, responding with medicines and curative or exorcising dances to G//amama's attacks or protecting themselves against anticipated onslaught by using prophylactic medicines. These tasks are part of the regular maintenance activities of the individual, the household, and the band. This defensive set is reversed in folktales. For example, when

the origin of the Okwa valley is related (outlined later and in Silberbauer, 1965:96), men and animals harass G//amama, inflicting on
him all sorts of injuries and indignities. G/wi derive much pleasure
and amusement from this reversal of circumstances.

The relationship between N!adima and G//amama is unclear.
Some informants believe that the two sometimes cooperate in doing
away with individuals of whom N!adima has "grown tired."
Whether this is by N!adima's design or by accident is not known.
There is some inconsistency in the idea that life is N!adima's to bestow but that G//amama might also be permitted to terminate life. In
other respects, N!adima is depicted as a jealous and possessive
being. A few informants expressed a vague concept of N!adima and
G//amama as two aspects of one being. Although consensus among
desert-dwelling G/wi rejects this idea, I encountered it among other
Bushmen.

When misfortune occurs, there is no certainty that it is caused by
the agency of either being. The G/wi have no processes of divina-
tion, and interpretation of this sort of incident is arrived at by discussion. The diagnostic characteristic of supernatural agency is that
the accident transcends normal experience in that it is not attribut-
able to common human error, to the usual operation of natural systems, nor to the customary behavior of animals. If the afflicted individual recovers or is rescued, then the agency is usually assumed to
be G//amama's; N!adima is believed to be unfailingly successful, and
if he does not cause the death of the unfortunate victim, it is because
he changed his mind at the last moment. The following incident illustrates the last type of situation. A small but highly venomous
buthicine scorpion entered the ear of a man who was digging up a
springhare. The scorpion could not be dislodged by any means available to the band; to disturb it would almost certainly cause it to sting
the man and kill him. To leave it in his ear would probably only post-
pone the dreadful moment. At the end of the day I returned earlier
than expected from a reprovisioning trip and was able to extract the
scorpion with a pair of tweezers. (The maneuver was hazardous
enough to incline me to share the band's view that N!adima had
changed his mind and intervened to help me remove the scorpion
before it could sting the man.)

Another incident illustrates the process of deciding whose agency
has caused misfortune. A party of men on an extended hunt away
from the band encampment were sitting at their fire late in the evening. Without warning, a lion attacked and grabbed one man by the
leg. He and his companions drove the lion off, but it attacked again,

55

concentrating on the same victim. The men then moved their camp several miles, taking their badly injured companion with them, and made a large fire at their new camp. The lion attacked once more that night and yet again in the following afternoon. This was a most unusual sequence of attacks, for lions seldom attack people, especially those who are awake, and, in any event, they first make their presence known by roaring and grunting. The hunters had not been successful, so their camp was without meat that might have attracted the lion. For the animal to be so persistent and return again and again was also exceptional. After commenting on these features, the men decided that N!adima must have sent the lion. After the lion's fourth attempt, they had no doubt that it was N!adima's hand that was against their companion, but at no stage did they abandon him or lessen their strenuous efforts to fight off the lion. I asked them if their interference would not make N!adima angry. "If N!adima hates //āũdze [the victim] he will kill him and not bother with us. He knows we will always fight the lion." When the lion did not return after the fourth attack and //āũdze did not die from his wounds, the men began to wonder if it had been N!adima after all. When //āũdze started to show signs of recovering from his injuries, they changed their minds and decided that G//amama must have sent the lion. A further point to note is that, although blood poisoning is known to be one of the ways in which both N!adima and G//amama may injure people, nobody expected G//amama to interfere with //āũdze's convalescence by complicating his wounds with sepsis. Having escaped being killed by the lion, he was thought to be safe, and he did, in fact, make a full recovery.

Despite the initial confusion as to which of the two entities wished to kill //āũdze, it was clear that N!adima and G//amama were considered to be separate and distinct and were not seen as differing aspects of the one entity. That the same means might be used to bring about the same end, namely, the death of a man, raises the question of why G//amama should be able to take life conferred by N!adima and regarded as belonging to him. This was beyond the informants' range of explanation; the G/wi believe that G//amama has lethal powers, which he will use unless frustrated by the various means that N!adima apparently made at the time of creation and that man has since had to discover for himself. This is consistent with the view of the universe as a systematic whole in which man and other creatures have to devise for themselves the optimal modus vivendi within the confines of the subsystems created by N!adima. The restraints are:

56

1. The creature's characteristic abilities and needs, ordained at the time of its creation
2. The obligation to coexist with other creatures, which arises out of their common status as N!adima's property to which due respect must be shown
3. The complex of interspecies and intraspecies dependencies
4. The operation of astronomical, climatic, biological, and other systems and subsystems
5. The pressure of condign sanctions "built in" by N!adima against wanton destruction or unseemly disturbance of his creatures and disrespect toward his creation and, by inference, to himself

There is, thus, no concept of man's primacy as the most favored among creatures. Instead, man has a place among them in a matrix of interdependent, interacting systems in which statuses are complementary rather than ranked. N!adima's universe has no objective other than to exist in the form of a self-regulating system. N!adima, in his remoteness from his creatures, is seen not as the embodiment of love nor of good, but as the one who created and ordained this self-regulating system, as the personification and source of order in the universe.

Man

In this uncertain universe in which man has to devise his own means of survival with no promise that a remote N!adima will intervene to help him, he is alone and such security as he has derives from the order that N!adima ordained and from his fellow men. The G/wi do not have lineages, nor do they have a strong sense of tradition or identify closely with their forefathers. Forefathers, or generalized ancestors, constitute a category, *g//onkhwena* (great, old people), who made most of the discoveries that the contemporary G/wi have as their stock of knowledge. There is, however, explicit recognition of fluctuations in the extent of that knowledge and of changes in G/wi culture. Much knowledge, for instance, is believed to have been lost in the smallpox epidemic of 1950 when many bands were decimated and dispersed. Some techniques, although remembered, have fallen into disuse (e.g., the use of pits in hunting giraffe – the pits still exist, but the practice has been abandoned). Since then, some old knowledge has been rediscovered and new added (e.g., the Iron Dance is now an alternative to the Gemsbok Dance, which was previously the only curative and exorcising dance). The physically dead are also socially dead and are not venerated. The knowledge that the *g//onkhwena* contributed is respected but is not held in any higher regard than its pragmatic, utilitarian value warrants, and tradition is not petrified and perpetuated by its elevation to sacredness. People do not feel themselves bound by the past but as being free to accept change and devise novel solutions to problems (e.g., the polyandrous solution described in Chapter 4). At the same time, recognizing the merits of known and proven solutions, the G/wi are conservative in their preference for these.

Like hunger and thirst or heat and cold, disease and illness are seen as environmental dangers that man must deal with in the best way he can devise and discover. The causes of diseases lie either in endogenous weaknesses in the patient or in his interaction with external factors in the environment or in a combination of these. N!adima and G//amama may direct the action of pathological agents in some cases, but most illness is seen as a random occurrence in which several factors, some beyond human knowledge, combine in a particular pattern of misfortune. It is known, for instance, that smallpox is an infectious disease and that to survive an attack is to gain immunity from future infection. Furthermore "smallpox is weak when it begins: people only start to die from it when many have fallen ill." But nobody knows why some contract the disease early in the epidemic and why others only do so later. Nor do they know why some are untouched by an epidemic (apparently without the protection of some previous postinfective immunity). The vernacular etiology of malaria links the disease with anopheline mosquitoes and recognizes that there is a 14-day incubation period but stops short of explaining why some who are bitten fail to develop any symptoms. Nor is there even speculation about why some wounds turn septic and lead to blood poisoning, whereas others heal. The difference is accepted with, respectively, stoicism and relief.

Endogenous weaknesses range from congenital abnormalities, which are rare, to the infirmity of old age. The latter is seen as the inevitable consequence of passing years. In only a few cases could explanations be found for birth defects (e.g., a small boy's mental retardation was ascribed to his mother's promiscuity, for I was told that the semen of so many different men had harmed the fetus). In most instances no causes could be found.

The G/wi have no sorcery, so this is not seen as a cause of illness.

Vernacular medical practices are not elaborate and the *materia medica* is not extensive. In combination they are, of course, adequate to the survival needs of the population, and the G/wi expect them to be effective in all but complicated cases of the conditions for which treatment is known. Treatment that fails unexpectedly is believed to be the work of N!adima "preventing the medicine" (denying it its usual power). Remedies are graded in their effectiveness – a minor remedy for a minor ailment, stronger measures for more serious problems. Medical treatment offered by my wife and myself was readily accepted and then sought not as a replacement of G/wi medicine but as a complement to it. Injections were regarded as the best remedy—"our medicine and your pills or drafts or ointments only treat the place where there is trouble. An injection goes through the whole body." Informants went on to explain that illnesses, even localized infections (excluding those of a minor nature), are systemic phenomena. "Even if the pain is in one place, the whole body is made bad by that thing which causes the pain."

Although medical knowledge is not esoteric, some men and women have a more extensive repertoire of cures or are more skilled in diagnosis and application of the correct remedy. Their advice is sought, and although there is nothing resembling specialist medical practice as an occupation, effective help with a health problem is repaid with a gift or favor as is the rendering of any other type of service.

Isolated by geography, the remoteness of N!adima, and the lack of close identification with their ancestors, the G/wi must find their security in cosmic order and the validation of their own identities and acts in the social order. The stark and lonely interdependence of the G/wi legitimizes, in the terms of Berger and Luckmann (1967), the positive value they place on the establishment and maintenance of harmonious relationships in their social order. The G/wi are not given to introspection and philosophizing and do not verbalize an ultimate rationale of their activities. However, from the emphasis placed on congenial company as an essential ingredient of satisfaction, the frequency and force of public and private reference to such company when enthusiastically appreciating pleasurable occasions, and the regret expressed when company is missing in otherwise happy circumstances, it is evident that the enjoyment of the company of friends and acquaintances is an important aim in G/wi life. In the absence of economic, social, and political stratification and the lack of religious ritual, which might otherwise provide alternative focal points of activities and values, and of rationale for action, harmonious relationships become the dominant theme in G/wi life.

The G/wi view of the nature of man is consistent with their desire for harmony in their social relationships. The noun stem *khwe-* can be translated as "man" in the general sense. Its use implies pos-

A young wife tells a story at her parents' fireside to an audience of family and friends. She and her husband have their hut behind that of her parents.

session of the qualities of *g//ahasi* (friendliness), *mahasi* (generosity), *≠anxasi* (wisdom), and (*!wamsi*) calmness and good humor. To lack these qualities is to be dubbed *khwemkjima'a* (man-not-is, i.e., inhuman). Known ethnic categories consist of people who are regarded as being inherently good, and there is the expectation of finding these qualities in them.

Although the G/wi are initially shy and reserved toward strangers, they anticipate the eventual development of good relationships with them and proceed on the basis that the strangers possess the desired human qualities. The display of unpleasant characteristics, such as *//xasi* (short, bad temper) or *xamxasi* (literally, lionlike, i.e., harsh, fierce, a violent temper), is considered an aberrant departure from intrinsic humanity rather than indicative of underlying evil. Such aberrations can and do become chronic. The response to persistent unacceptable behavior, which is described in Chapter 4, does not involve rejection of the miscreant as a *khwema* (man) but nevertheless ensures that he does not remain in any band long enough to cause lasting or serious disruption. In their initiation school, boys and young men are told, "Even if you visit very far among *khwena* [people, here specifically Bushmen] whom you do not know, you must not keep to yourself and ignore them. Even if a *xajekhwema* [stranger met for the first time, literally, entering-man] comes from very far where nobody knows him [i.e., you have no common acquaintances], do not ignore him. Go out and be friendly. Open yourself to all, for all *khwena* are good. Do not fight them, for they will never fight you." (The injunction "open yourself" is akin to "unto thyself be true" rather than to a candid self-relevation to all comers.)

As generalizations of their expectation of experience with other human beings, this idealistic model of man really only applies to other G/wi. Non-G/wi strangers are certainly not ignored, but the cautious reserve with which they are initially treated is hardly in the spirit of the instructions given to initiands. The G/wi stereotypes of ethnic groups are also at variance with the idealistic model of man. The most familiar ethnic group among Bushmen are the G//ana, whom the G/wi seem to like well enough but regard as being inclined to *sehasi* (greed). The Nharo (who are more correctly called by their self-appellation, G≠eikhwena, People of the Calcrete Outcrop, but Nharo, the ≠āũ//ei name for them, is current in the literature) are said to be angry and likely to help themselves to the possessions of the G/wi. The pejorative nature of the stereotype increases with distance and rarity of interaction. The ≠āũ//ei are greatly feared for their supposed fierceness and ability to transform themselves into lions and prey upon people. The G//ulu are believed to be cannibals with voracious appetites. Non-Bushmen are categorized as ≠*ibina* (Bantu in general but usually Kgalagari, with whom the desert G/wi have occasional contact) or *lo:xana* (Europeans). Further divisions into tribes or nationalities do not exist and I saw no evidence that the G/wi have any cognizance of these. Although the names are not pejorative, they lack the favorable connotations of *khwena*, which is used only in reference to individual Bantu and Europeans when they have become known and trusted. The stereotypical non-Bushman is short-tempered, aggressive, and given to irrational actions. Nevertheless, Bantu and Europeans are considered to be superior because of their apparently greater knowledge and physical strength. (The G/wi consider themselves inferior to all other peoples in the extent of the progress they have made in discovering the means and techniques of overcoming the problems besetting man as one of N!adima's creatures.) Although afraid of non-Bushmen, they do not consider them to be evil but as having dangerous or potentially dangerous ways of behaving, which one must first learn to avoid or deal with before establishing friendly relations.

The category of human beings is concentrically subdivided, the radial distance of the subcategories from Ego being inversely propor-

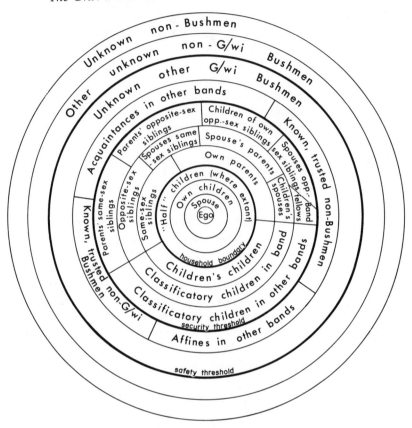

Figure 7. Categorization of human beings in the G/wi culture.

tional to the frequency of interaction and the intensity of emotional attachment between Ego and those in the subcategory. In the subcategories represented by the three outer rings of Figure 7, radial distance is a function of perceived ethnic distance and is a projection of the categorization of attitudes toward known individuals in those ethnic categories. The contents of the inner-ring subcategories is determined by the kinship taxonomy; the innermost two or three rings represent the household, the personnel of which vary with age and the status of age. In the case of unmarried individuals, the inner ring consists of same-sex siblings (with whom interaction is most frequent and emotionally intense), with parents and opposite-sex siblings being placed in the second ring. The figure represents the situation of a mature individual, that is, a married man or woman with married children. Although the rate of interaction is approximately uniform around each ring, the extrahousehold rings are divided into segments, as the quality differs, for example, according to whether there is a joking or avoidance relationship between Ego and the persons concerned.

The heavily outlined rings in Figure 7 represent, respectively, the household boundary, the security threshold, and the safety threshold. The security threshold contains categories of persons with whom relationships are normally predictable and reliable and in which any element of threat is expected to be contained and eventually resolved by mechanisms of social control. (The normative aspects of the kinship system are binding on all persons within this circle.) The safety threshold includes all persons with whom relationships are not expected to contain unforeseen threat or danger. Beyond this threshold, relationships will not necessarily involve danger, but an element of unknown, hence unpredictable, risk is present.

Fauna

The animals, like man, are N!adima's creatures; as his property, they must be respected, not abused. They may be killed in self-defense or for food or to avoid an attack that is believed to be imminent. When playing at hunting, children are allowed to kill dung beetles and other harmless, inedible creatures, which they fantasize as prey animals. If a child should kill too many, however, he is chastised for being a greedy hunter, and behind such admonition there appears to be a genuine concern that such behavior by children will anger N!adima.

Humans are distinguished from other animals by the obvious differences in appearance and habits and by man's unique possession of a *g/ama* (postmortem spirit). There is an underlying enmity inherent in man's position on the trophic web: "we hate the lions, leopards, and spotted hyenas because they will hurt us. The antelope [listed by name] hate us because they see our fires at night and N!adima has told them that these fires are to cook them." The comparable affective models of the relationships existing at intra- and inter-specific levels among the fauna reflect the ecological relationships of predation, competition, and protocooperation. (In their ordering of these relationships, the G/wi see parasitism as a type of predation; commensalism and mutualism are interpreted as protocooperation, and neutralism is not significant to them.) Such relationships and the quality of interaction to which they give rise are part of the natural order ordained by N!adima.

The affective nature of the models is a consequence of the anthropomorphic nature of G/wi ethology in which animal behavior is perceived as rational and purposive and directed by motives based on values that are either held by the G/wi themselves or by other peoples known to them or are negations of such values. The motivational and value systems of animals are not isomorphic in all respects with G/wi and other human systems but are empathetically modified to fit the perceived circumstances of the animals themselves. Each species is credited with characteristic behavior, which is governed by its *kxodzi* (customs), and each has its particular *kxwisa* (speech, language). Protocooperating (i.e., mutually beneficial) species and even some that are hostile to one another can understand one another's language, and some animals can even understand a certain amount of G/wi. Man, too, can understand a limited amount of the speech of some species, for example, the alarm cries of birds. Baboons, the most versatile of animal polyglots, eavesdrop on G/wi hunters and pass on the plans of the hunters to the intended prey animals. This is not altruism but is caused by the baboons' legendary love of trickery and teasing.

The special capabilities of some animals are believed to have been arrived at by rational thought and then institutionalized as elements of the species' *kxodzi* (customs) after having been passed on by the discoverers or inventors in that population. (For instance, the ability of the penduline tit, *Anthoscopus minutus*, to manufacture feltlike nesting material from the awns of grasses.) Such capabilities are compared with those that man has developed in the process of devising means of meeting environmental pressures. Some species possess knowledge that transcends that of man; the bateleur eagle, *Terathopius ecaudatus*, for instance, knows when a hunter is to be successful and hovers above him, thus acting as an omen of sure success (but not, apparently, as a warning to the prey). Several animals are believed to be able to foretell the extent of wet-season rains and the location of the best falls and to plan their annual behavior cycle accordingly. These animals are seen as having rather critical limits of tolerable error. If they are to reproduce successfully, they must time their activities to gain the greatest benefit from the rains when they do come. They are believed to have a more sensitive perception of how the rains are developing and can thus furnish the discerning observer with more accurate information than his own less finely tuned senses can gather. The duiker, *Sylvicapra grimmia*, practices sorcery against his animal enemies and even against conspecific rivals, and some steenbok, *Raphicerus campestris,* are thought to possess a magical means of protecting themselves from a hunter's arrows.

A few animals (e.g., the crimson, velvety red earth mite) are said to be N!adima's children. These are distinct from the children born to N!adisa in the sky country and are recognized by their striking appearance and what are regarded as their privileged existence, that is, not having to expend any effort in finding their food and in defending themselves against enemies. They are harmless and small and are neither feared nor hunted. Although neighboring Bushmen and Bantu have quite a menagerie of mythical beasts, the G/wi have none (the underworld monsters are excepted from this discussion as they belong to another region of the universe).

The behavior of all of the animals in the G/wi bestiary is seen as bound by N!adima's order, can be accounted for in terms of known or knowable regularities, and is believed to be rational and directed by intelligence. Despite its anthropomorphic basis, G/wi ethology accounts for animal behavior with sufficient accuracy to make it an efficient aid in planning hunting tactics, in anticipating the actions of dangerous creatures, and in interpreting the connotations of interspecific relationships. A large part of the spectrum of available environmental information is habitually scanned and integrated to form a detailed picture of the behavior of a wide range of species from one hour to the next. There are some puzzling errors, however, for example, that steenbok have a regular breeding season, when, in fact, these small antelope drop their fauns year round. The anthropomorphic ethological models have nevertheless withstood the rigorous empirical testing that their use imposes and, although so at variance with those of European ethologists (e.g., Eibl-Eibesfeldt, 1970), furnish an interesting confirmation of the point made by Wallace (1961), Berger and Luckmann (1967), Goffman (1959), and others that conflicting cognitive maps can serve as equally effective equivalents up to a certain level of analysis.

Mammals

G/wi folk knowledge contains more information about the ethology, anatomy, and physiology of mammals than about the other zoological classes of central Kalahari fauna. The concepts of mammalian ethology are more overtly anthropomorphic than are the concepts concerning other classes. As a class, mammals feature more prominently in conversation and have a higher priority in environmental description than do other animals. From this I conclude that the G/wi regard them as being closer to man than are other vertebrates or members of other phyla.

The taxonomy is covert in that there is no vernacular term for mammals. The criteria of classification are tetrapedal gait, warm blood, fur, and mammae or external male genitalia. Paradoxically, the nearest generic term, *kx'o:xudzi* (literally, to-be-eaten-things, prey), includes mammals that are not hunted, for example, lions and hyenas, but excludes birds, reptiles, batrachians, reptiles, and insects, which are regularly included in the diet. These latter, along with mammals, are *khweahasxudzi* (living-things, creatures).

Specific mammal names reveal a tendency common to the G/wi nomenclature of all life-forms, namely, that monomorphemic, specific names have been given to those that are important because of their frequent proximity to man, their contribution or threat to man's survival, or their remarkable habits or appearance. Life-forms of lesser importance usually have polymorphemic, derived names or generic terms to cover several species.

Each of the large mammal species is a prey species, is dangerous to man, or competes with him for prey animals and other resources, and each has a specific name:

> *djuama* (kudu, *Tragelaphus strepsiceros*)
> *g!eima* (steenbok, *Raphicerus campestris*)
> *xamma* (lion, *Panthera leo*)
> *!'wema* (leopard, *Panthera pardus*)
> *g//aduma* (Cape hunting dog, *Lycaon pictus*)
> *//'ama* (bat-eared fox, *Otocyon megalotis*)

Many small mammals are seen as peripheral to man's interests and are either covered by generic terms or have derived names:

> *n!unima* (rat, mouse, gerbil, dormouse)
> *lxo:g/wema* (literally, gemsbok-mongoose, ground squirrel, *Xerus inaurus*. The G/wi name reflects a fancied resemblance between its lateral stripe and the belly line of a gemsbok)

Generic grouping is preponderantly phenetic, although behavioral characteristics are taken into account; for example, *damama* (polecat, *Ictonyx striatus*) is not included among the mongoose (each of which also has a specific name), although it is phenetically closer to them than is the ground squirrel. Its essentially nocturnal habit, solitary nature, and defensive use of its anal gland are evidently behavioral characteristics that earn it its distinctive taxon.

The anthropomorphism in G/wi ethology extends to the terms

used in interpreting mammal behavior. The prey animal, by using its strength and wits in the manner characteristic of its species, is expected to do its best to avoid the hunter. He, in turn, expects to overcome the difficulties of the hunt, the animal's behavior and other circumstances, that challenge his own skill and cunning. Also, there is always the chance of being outwitted or otherwise frustrated by the individual animal's idiosyncratic quirks. A gemsbok that uncharacteristically doubled back on its tracks and thus avoided an ambush annoyed the hunters at the time, but later they expressed admiration of its being /xudi (ingenious – used of a person who devises a novel and effective solution to a problem). However, a similar incident led to the hunter's describing a hartebeest as /xan (useless, stupid) for not conforming to //xamakxodzi (hartebeest customs). When watching a herd to select their target, hunters classify individual animals by terms used for human attributes of personality and character, rejecting those animals that are judged to present too much difficulty because of their contrariness, courage, and so on. Some are ka: (cheating) or !ao (cowardly); others are ≠'u: (insolent) or khwa:khwa (conceited) and may therefore be likely targets. The typology is apparently exhaustive and contains more than the 18 categories that I recorded. It provides the basis for predicting the animal's behavior before and after it has been shot. (G/wi arrow poison takes some time to weaken the animal sufficiently to allow the hunter to come in for the kill; until then, the wounded prey must be followed at a distance.) In my observation, the diagnosis usually proved to be an accurate prediction of the animal's behavior once the hunt commenced. Although the hunters obviously knew the signs and connotations of each type and state and tried often enough to explain them to me, I was seldom able to distinguish diagnostic characteristics.

The G/wi also project their own values and habits in explaining other types of mammal behavior. The gregarious species of antelope, for instance, are believed to discuss their migrations in much the same way as does a G/wi band, and the "nursery" herds of springbok are seen as a parallel to the G/wi custom of leaving the children in the care of a campkeeper during the day. The doings of lion families are explained in terms of behavior typical in a G/wi household, and contrary to experts' descriptions of lion behavior (e.g., Guggisberg, 1961; Schaller, 1972), the male is believed to do most of the hunting.

Mammalian anatomical terms and physiological processes are equated with those of man. In the course of gutting a kill, I have been given the name of each organ of the prey animal, together with an account of its function. Where appropriate, the position of the human homologue was indicated and its matching functions described. (I was thoroughly instructed. The lessons were very popular occasions when girls and women were within hearing and could be teased by my informants, who gave luridly bawdy versions of the parallels in, particularly, the ethology, physiology, and anatomy of reproduction.) Organs are linked in conceptualized systems, for example, the cardiovascular-respiratory system, and the G/wi accounts of these systems matched fairly closely those of European physiologists. There are, however, some significant differences. In the G/wi version, for instance, blood flows away from the heart through the veins and returns via the arteries. Semilunate vascular valves are seen as pumps whose action is auxiliary to that of the heart. The G/wi physiology of digestion includes a concept of nutrient extraction and circulation; the darker hue of venous blood was explained in terms of its nutrient load, and arterial blood, lighter in color, had shed its nutrients and returned empty to the heart, liver, and lungs to collect a fresh supply. In this regard it is noteworthy that man is described as monogastric, in contrast to the four-chambered structure of the stomachs of ruminants, which are the mammals most commonly dissected by the G/wi. Laughlin (1962) has ascribed the extensive anatomical knowledge of Manchurian Tungus and of Aleutian Islanders to their practice of postmortem human dissection combined with observation of other mammals in their environment. From Laughlin's description, these peoples' knowledge appears to be no greater than that of the G/wi, who do not dissect their dead and have no opportunity of which I know of observing man's internal anatomy directly. By contrast, P. A. Silberbauer (pers. comm. 1969) found the Pitjantjatjara of the Musgrave Ranges, South Australia (who, like the G/wi, do not practice human dissection), to have a comparatively scanty and inaccurate knowledge of human physiology and internal anatomy.

The G/wi account of the reproductive processes in man is essentially in accord with the views of European physiologists. It is only in this field that a terminological distinction is made between man and other mammals; at the end of her pregnancy a woman will *aba* (literally, to carry on the back, as a baby is carried on the back; to give birth), but other female mammals *g//am* (calve, whelp, farrow, etc., to give birth).

The relationships between the G/wi and the larger mammals is of considerable economic importance as the bulk of the meat intake is derived from hunting larger antelope and the greatest perceived danger is from the larger predators. Consistent with this situation is a keen awareness of the order that N!adima imposed on these relationships. I have already mentioned the strong disapproval expressed by the G/wi for what is seen as greedy hunting, that is, shooting more than immediate needs require. Similar disapproval is voiced when lions enter G/wi camps. When the lions pestered us for 18 nights, for example, much indignant comment was expressed on the impropriety of the lions' behavior and the consequences that would surely befall them, as N!adima would be angered by their insolence in contravening his regulation of the relationships between lion and man. The perception of, and concern for, the obligations of life-forms other than mammals are less acute.

Birds

Although the G/wi are keen observers of their environment and have a considerable knowledge of bird behavior, certain gaps in their bird lore indicate that the avifauna is less closely observed and less well known than are the mammals. This lack reflects the fact that birds are of lesser importance to G/wi subsistence and in other aspects of their lives.

The generic term, *dzeradzi* (birds), covers the whole avifauna. (However, it would be regarded as amusingly pedantic to refer to an ostrich, *g/edoma*, as a *dzerama*, bird.) The lesser importance of birds, compared with mammals, is also indicated by the greater incidence of generic and derived names for birds. Generic grouping is phenetic, for example, *g/ū:dzi* (owls) and *kx'keidzi* (the larger diurnal raptors, including falcons), most species of which also have monomorphemic or derived names:

> *n≠a:tsoma* (secretary bird, *Sagittarius serpentarius*)
> *g!we:≠ubima* (literally, clatter-wing, bateleur eagle, *Terathopius ecaudatus*)

(I was disappointed by what initially appeared to be a glaring error in G/wi bird lore, namely, the statement "No man ever sees the nest of the secretary bird." However, Smithers [1964] states that it is unrecorded in the central Kalahari. It does, in fact, occur sporadically and is probably a nonbreeding migrant, which explains why the G/wi would not have seen the large and obvious platformlike nests built by these largely terrestrial eagles.)

Specific nonderived names are given to many birds that are of no

economic importance but are nevertheless either very common or
have unusual habits:

> *n//ajisa* (nightjar, *Caprimulga europaeus, C. tristigmata*)
> *dzausa* (Namaqua dove, *Oena capensis*)
> *gu'osa* (lesser striped swallow, *Hirundo abyssinica*)
> *n≠a:n≠edasa* (European swift, *Apus apus*)

(The last-mentioned name may originally have been derived, evidently from the perfect-stative tense of a reduplicated verb, but the
meaning is lost and the name is now specific.)

The majority of birds, however, have derived names. Many of
these are onomatopoeic, being echoic of the bird's call:

> *ki:hohoma* (red-faced mousebird, *Urocolius indicus;* has a three-note
> call, the first note slightly higher in pitch than the others)
> *dwē:sa* (ant-eating chat, *Myrmecocichla formicivora,* has a short,
> plaintive, slightly nasal call, which the G/wi render as *dwē:*)

Other derived names are descriptive of some aspect of the bird's behavior. An example is *!udo/osa* (cardinal woodpecker, *Dendropicos fu-
scescens.* The name means "knocking-feather," after the ringing
sound of the bird hammering its beak into wood in search of food
and when excavating a nesting hole in a tree.

"Clatter-wing," the bateleur eagle, has already been referred to.
The observation of both habit and sound are combined in the name
of the red-eyed bulbul, *Pycnonotus nigricans,* an aggressive species
with a call perfectly matched by *≠wig!wirida≠xeisa* (red-angry-it-I-
eye; i.e., it is I, [with] the angry red eye).

Most of the birds regularly taken for food have nonderived, specific names:

> *g/edoma* (ostrich, *Struthio camelus*)
> *g/euma* (kgori bustard, *Otis kori*)
> */xanima* (crowned guinea fowl, *Numida meleagris*)

The explanation of bird behavior is as anthropomorphic in character as is mammalian ethology. The seasonality of breeding is interpreted as the birds' rational foresight in timing the hatching of their
eggs to coincide with a plentiful supply of their particular foods
(which respected wisdom the G/wi themselves do not imitate, bearing as many children in lean, as in flush, seasons). A quite human
propensity for error is attributed to colonies of *g/wane:dzadzi* masked
weavers (*Ploceus velatus*), which occasionally "miscalculate" the
quality of a forthcoming season and then abandon their nests and
eggs, and even young, when it fails.

The change in breeding plumage and the courtship displays of many species are seen as amusing instances of male vanity. Distraction display is seen as indicative of cunning and courage in the parent birds, but the ostrich, which often advertises the presence of its young by premature distraction display, is derided as a fool that unsuccessfully imitates somebody else's trick.

Nest parasitism by the several species of cuckoo is mildly deprecated but not regarded as being very wicked, as eggs are not considered to be alive until just before hatching, when N!adima infuses life into them. (The "proof" of this contention is the fact that many birds produce a fresh clutch of eggs if the hatching process of the first clutch is interrupted or excessively disturbed.) The parasitizing cuckoo is not, therefore, considered to evict the "children" of the host pair. Cuckoos provide mild drama for the G/wi. Watching the host nest, the cuckoos wait for the parent birds to leave to search for food. The hen cuckoo then enters the nest, lays an egg, and throws one of the hosts' eggs out. Meanwhile, the male cuckoo keeps watch and warns his mate of the return of the hosts. The warning call of the diedrik cuckoo (*Chrysococcyx caprius*) is rendered by the G/wi as "*Be:, be:, be:, akhwoda ha:*" (Flee, flee, flee, here they come, he and she). Parasitism by the shaft-tailed whydah (*Vidua regia*) is viewed quite differently; the male bird which assumes a magnificent black-and-yellow breeding plumage and grows a tail three to four times the length of his body, has a harem of 20 or more hens. This bird is said to be one of N!adima's earthly children (not to be confused with those in the sky country), which is why it leads what is seen as a particularly privileged life.

Vultures are believed to live in bands, each of which is ruled over by a *//eixama* (chief, headman), the lappet-faced vulture, *Torgos tracheliotus*. An aggressive and rather solitary species, this vulture is less common in the central Kalahari than are other vultures; because of the lower density of distribution, it usually arrives at a kill after the commoner, more numerous other vultures have settled and proceeds to chase them off to make a place for itself. This behavior is interpreted as being similar to the perceived authoritarian nature of the role of headman or chief among Kgalagari and Tswana peoples.

I have referred to the special knowledge that some birds are supposed to possess. Swallows hawking low-flying insects in humid weather often presage coming rain; *gu'osa*, the lesser striped swallow, is particularly bold, sweeping across the ground to within a foot of people in the camp in order to inform them, in G/wi belief, that rain is coming soon. The annual migration of swallows and swifts is said to be guided by their long-term weather forecasts, and the G/wi presume that they move into some other country where there is water in winter, as the gregarious species of antelope and other migratory animals are believed to do.

The behavior of birds is of practical value as an indicator of situations that are of importance to the G/wi, for example, alarm cries and distraction display betray the presence of snakes, sudden changes in the normal search and feeding pattern may reveal the presence of some other animal in a thicket, and a pool of water can be detected at several miles' distance by observing the flight patterns of some birds. Knowledge of both normal behavior and significant deviations from those patterns is needed, and the doings of birds make up one part of the environmental information spectrum, which is under constant, if automatic, observation. Birds are considered to be intelligent creatures, free of many problems that beset less mobile animals. This freedom is seen as being manifested in a neutral, or somewhat friendly, attitude toward most other animals, including man. Credited with thought processes and values comparable with those of man, birds are thought to react to many situations in the way man would, and their behavior is therefore seen as having some of the value that human actions would have as a source of information.

Chelonians

Tortoises and terrapins form an unnamed, covert class. The existence of the taxon is revealed by informants' likening its members to one another, for example, that *g//udema* (hinged tortoise, *Kinixys* sp.) is "like" *g//we:ma* (geometric tortoise, *Psammobates* sp.). Tortoises are economically important as a source of meat in all except the winter months (when they hibernate) and their shells are extensively used as ladles and bowls. Each of the three species of tortoise and the one terrapin have specific, nonderived names. All are regarded as unintelligent creatures of proverbial hardiness. Some informants alluded to their antiquity, an interesting suggestion that the G/wi might have a concept of a prehuman era, but I was unfortunately unable to uncover further evidence of such a notion.

Ophidians and legless squamata

Snakes, legless skinks, and Amphisbaena are all termed */xaudzi*. Although only a few of the 18 or more species of snakes are venomous, all */xaudzi* are regarded as dangerous in some degree. The majority of the members of this taxon are named. Those that are common or are prized as delicacies have nonderived names:

> ng!amma (tiger snake, *Telescopus semiannulatus*)
> g//ajema (puffadder, *Bitis arietans*, also horned adder, *B. caudalis*, er-
> roneously believed to be the juvenile form of *B. arietans*)

Less commonly encountered species have derived names:

> g//omg//oma (literally, big-big, mole snake, *Pseudaspis cana*)
> g!utakxaima (literally, yellow-spitter, Cape cobra, *Naja nivea*)

N!adima gave snakes a hatred of man, which is why they attack man and cause severe stomachache in any young people who eat them. However, this does not justify killing snakes; they can be killed only in self-defense or to provide delicacies for old men and women who, "by reason of age and infirmity, are immune to the stomachaches." N!adima also created a small number of plant species that provide remedies for snakebite. I was unable to identify the plants from the fragmentary specimens provided (all obtained at third-or fourth-hand from those who had gathered the plants), nor was I able to judge their efficacy; on the very rare occasions when someone was bitten, there was always some unfortunate reason why the remedy could not be applied – the supply was exhausted, the snake had bitten in the night and the stock of medicine could not be found in the dark, and so on. Although there is a great fear of the consequences of snakebite, habitual watchfulness makes the probability of an accident remote. Yet watchfulness is not always all that it might be. My wife and I were astonished to see a fork-marked sand snake slither up to a group of girls lying on the ground, talking. The snake crawled across the back of one girl, and then continued on its way, unnoticed by any of them. It was about 1 meter long and the species (*Psammophis leightoni trinasalis*) is unpleasantly, but not lethally, venomous. Its appearance in a G/wi camp would normally provoke considerable agitation.

Legged squamata

The legged skinks, agamas, lizards, geckos, and chameleons form a vague, covert taxon based on the squamatous body plan and the characteristic permanently "open" eyes. Although the taxon is not named, its existence is indicated by informants' describing members of the taxon in terms of their perceived resemblance to reptiles, particularly the typical reptilian waddling gait, which informants imi-

73

tated. Fewer than a quarter of the species that I identified have G/wi names and most of them are virtually ignored. Some are hunted for their meat and a few considered dangerous to man. These have specific, nonderived names:

> *suduma* (white-throated leguaan, *Varanus exanthematicus albigularis,* which is hunted)
>
> *n!aduma* (chameleon, *Chamaleo dilepis,* believed to blow out some sort of venom when it hisses defensively – the poison causes "withering and death." Chameleons are protected by N!adima, who will "drive away the rain" if one is killed by man; see later in this chapter, under "Weather and the seasons")

Batrachians

Frogs and toads are grouped together by the generic term *ng!wabe:dzi.* Only the bullfrog (*Rana* sp.), which is a delicacy, is specifically named (*g!wema*). Tadpoles, *khwonekhwenidzi,* are recognized as the intermediate stage in the development of frogs and toads, and folk knowledge contains fairly detailed accounts of the feeding, breeding, and hibernating habits of many of the species, even those that are unnamed.

Invertebrates

The central Kalahari has a large and richly diverse invertebrate population. Only a small proportion is known to the G/wi and by no means all of the known species are named. Folk knowledge recognizes that invertebrates have exoskeletons, are generally oviparous, and pass through successive metamorphoses in their life cycles and to this extent they are grouped in a covert, unnamed class. The vernacular taxonomy of this class is overwhelmingly generic, with few individually named species. The classification is phenetic and, to some extent, coincides with the taxonomy of European entomologists at the level of orders. In general, the named classes are those containing members that are useful (providing food, arrow poison, medicines, or the means of decoration), those that are dangerous (i.e., which sting, bite, exude irritant fluids, or are believed to be vectors of disease), those that are particularly striking in appearance, and those that are nuisances:

> /inidzi (grasshoppers and crickets)
> g//amdzi (mantes, bugs, and beetles)
> g/ha:xamag//wa (literally, large-ant lions, ant lions)

gje:kibidzi (butterflies and moths)

g/enidzi (flies)

g//we ≠xei/xiowedzi (literally, geometric-tortoise-eye-sting-broken, wasps – i.e., their stings are so fierce as to penetrate even the eye of the proverbially tough tortoise)

gjine:dzi (bees)

≠wam ≠wamdzi (small ants)

g/ha:dzi (large ants)

n//abokije ≠we ≠wedidzi (literally, sandal-goes-scuttle-scuttle, wood lice)

g!õg!õnidzi (millipedes)

n!hada//eidzi (centipedes)

//xadidzi (scorpions)

tsa:tsam!adzi (literally, testicle seekers, solifuges or sun spiders)

g/uba:dzi (spiders and harvestmen)

tamminidzi (mites and ticks)

Other taxa are termites, lice and fleas, dragonflies, mosquitoes, and caddis flies. Within these taxa are individually named species, those of great importance having nonderived names (e.g., *//ua:aa, Diamphidia simplex,* the beetle from whose larva arrow poison is derived), whereas species of lesser importance usually have derived names (e.g., *dju:n!odzi,* literally, eland-caterpillars, and edible caterpillar).

The protection that covers N!adima's creatures extends to invertebrates, which may be killed only in self-defense or for use. Few species are considered to be innately antagonistic to man. Scorpions, which present some risk to the unwary, are only killed if their burrows are located where a shelter is to be made. It is said that there is no purpose served by simply removing them, as they invariably return to the site of the burrow and, enraged by the disturbance, will surely sting somebody. Jerrymanders, or sun spiders (*Solpuga caffra*), on the other hand, are not killed, although it is believed that when one of these wide-ranging solifuges finds itself among a group of men it will attempt to bite their testicles (hence the G/wi name). Its enormous jaws with their powerful, sharp teeth and the speed with which the sun spider rushes erratically over the sand are indeed daunting, but the danger it presents is not grave. Although the arrival of one is greeted with shouts of alarm and general male agitation, it is not killed but is deftly scooped up in the hand and thrown some distance away. The G/wi share with many Khoikhoi (see Schapera, 1930) the belief that millipedes have the habit of entering the ears of sleeping people and boring through to the brain to kill the sleepers. Nevertheless, millipedes found wandering around a shelter during the day are simply thrown away; only those that are seen at night are killed in the fire.

G/wi bee lore is curiously extensive, considering that the G/wi rob nests infrequently and that the notorious aggressiveness of the African honeybee (*Apis mellifera adansoni*) scarcely encourages close observation. We have an apiary on our small farm in Victoria and are amazed at the depth and accuracy of G/wi knowledge. We also profoundly admire the physical and intellectual courage of those who, protected by only a handful of smoldering grass, stopped to note the details of the organization of the colony they were busily robbing. It shows something of the nature of G/wi curiosity that these observations were supplemented by their watching the regular behavior of the species to form a knowledge of bees that went far beyond their practical requirements and would compare favorably with that of the average enthusiastic apiarist (who, remember, works in far greater comfort and has access to a splendid literature).

Knowledge of the physiology, ethology, and ecology of some other invertebrate species is quite detailed in relation to the species' interaction with man. Although the life-style of the species is conceived of as a system, detailed knowledge consists mostly of short, linear sequences of information, which seldom intersect to form systematic accounts. Knowledge of the species beyond the field of its interaction with man is sketchy. To exemplify: In its larval form *Diamphidia simplex*, the arrow-poison beetle, is perhaps the most closely studied invertebrate species. The larvae feed on the leaves of the corkwood shrub, *Commiphora pyracanthoides*, and then migrate down into the sand within the shrub's drip line to pupate. The feeding larvae are watched and the cocoons are dug up shortly after migration. This part of the life cycle is closely observed and well known, and questioning elicited the information that there are also egg and adult phases. Knowledge of, and interest in, these later stages are, however, markedly less, to the extent that informants have quite erroneously indicated to me a species of shield bug as the adult form of the beetle. As far as I could discover, G/wi bowmen do not look for eggs on the plant in order to forecast which corkwoods are most likely to eventually yield a supply of arrow poison.

This fragmentary knowledge contrasts with the extensive folk knowledge of the lives of mammals, all aspects of which are covered in approximately uniform detail and conceptualized as interdependent, with information concerning one part of a mammal's life-style being used as a basis for making statements about some other aspect (e.g., observation of ecological variables to provide a basis for prediction of behavioral tendencies). The extent and uniformity of knowledge and its systematic nature progressively decline in relation

to birds, reptiles, and batrachians and are least in relation to invertebrates, and the extent to which human characteristics are attributed
(or the extent to which characteristics are shared by animal and man)
is matched by the breadth of the spectrum of folk knowledge of the
class.

This scale of knowledge and of characteristics shared with man is
consistent with a scale of conceptualized distance from man, comparable with the egocentric conceptualization represented in Figure 7.
It may be similarly represented by a series of concentric circles with
man in the center, mammals occupying the next ring, after which
come birds, chelonians, and so on, with invertebrates, conceived of
as the animals farthest from man, placed in the outer ring.

The flora

Plants are clearly included among life-forms, for they are said to *khwe*
(live), *//o:*(perish), *!xu:* (reproduce), *g//o:* (grow), and *be:* (fear, be afflicted by) the cold of winter and the heat of the summer sun. However, *ji:dzi* (flora in general, trees in particular) have no power of lo-
comotion, do not feel pain or experience pleasure, and have neither
will nor intelligence. Their patterns of life were ordained by
N!adima when he created each species, and although he can make a
plant live or die, he cannot otherwise vary the set pattern of its life to
make of the plant an instrument of his will. The power that a species
has to heal, nurture, or harm man and animal is inherent in every
specimen of that species and is automatically and mechanically released when the plant is used or eaten unless N!adima intervenes to
"prevent the medicine."

The absence of mobility and volition sets the flora apart from the
animal life-forms. To the G/wi, this makes them an integral, but variable, aspect of the land to the extent that vegetational cover is one of
the criteria in the classification of topography (see later in this chapter, under "The Land"). Conceptually, the flora links the vital dynamic aspects of the land with its unchanging, static fabric.

Some species are seen as *!an* (useless) and others are utilized by
man in competition with animals. It has not escaped G/wi attention
that some "useless" species are eaten by animals on which they
themselves feed; the epithet implies the absence of immediate utility
and does not constitute a denial of the concept of a trophic web, of
which the G/wi have a clear perception. Nevertheless, the purpose
behind the creation of the whole floristic spectrum is not known. No
plant is thought of as having been created for the special benefit of

anything else, and neither man nor any other creature has preferential rights to any species.

As N!adima's property, plants are protected by the same prohibition of wanton destruction as are faunal species. All creatures are permitted the use of plants in order to live, but they may not waste them. Apart from their fear of offending N!adima and bringing upon their heads his punishment, the G!wi avoid stripping an area of a species. Exploitation ceases when the supply shrinks and a residue is left, so that regeneration is not imperiled. Informants described this as a conservation measure. The extra work entailed in searching for scarce specimens must discourage exploitation beyond a certain point; nevertheless, the conservation intent is demonstrated by their refusal to use locally scarce specimens even when these are found while gathering other species.

Despite their perception of conservation principles and their expertise in field botany (see later), the G!wi have only slight, and often inaccurate, knowledge of plant physiology. Plant reproduction is not believed to be sexual. Although a flower has male and female components (the male parts being the caducous portions, that is, petals, etc., which drop off, and the female parts those that remain on the plant and grow into seed) and the process of reproduction is seen to be analogous with animal reproduction in that the parent organism manufactures the substance and form of the offspring, the G!wi interpretation does not include fertilization of female parts by male parts. When the seed ripens it falls to the ground and waits in the sand for the rain. Rain is the agent by which N!adima vitalizes the seed, which then starts growing. It is recognized that seeds require suitable environments and that not all seeds will grow. It is also thought that germination is inhibited if "the land is angered" (i.e., if the subterranean spirits, or even N!adima himself, are offended) by the breaking of some taboo and that the land may "refuse to allow the seeds to grow." N!adima may, of course, also express his displeasure by withholding rain, which is a more severe sanction, involving the loss of not only ephemerals but perhaps also perennials.

Rain is also seen as the food of plants. The main function of roots is to hold the plant up, but they also drink for the plant. Leaves are there to shade the roots and protect them from the heat of the sun; and the "proof" of this is that the roots are killed and the plant dies and falls over when the leaves fall in very hot weather. (This is not always the case, as leaf dropping is a stress-avoiding mechanism, but it does happen often enough that trees and shrubs die from drought.) As further "proof," after autumn, when the sun's heat is no longer dangerous, deciduous species (the majority of perennials in the central Kalahari) drop their leaves because there is no further need to shade the roots.

The periodic activities of plants, their flushing, flowering, and seeding, are timed by N!adima's ordination so that his creatures will have at least some food in all seasons. The witgat (*Boscia albitrunca*), for example, bears its sweet juicy berries before the onset of the rains. It is one of the few foods available at this time but is usually sufficiently plentiful to sustain a concentrated population and thus permits dispersed Glwi bands to re-form earlier in the summer than would otherwise be possible. Informants rejected the suggestion that N!adima might have had this particular fact in mind when ordaining the time of fruiting. In their view, N!adima is only concerned with survival of his creatures; the quality of life is their own concern. Had man not the wit to realize that witgat berries permitted a longer phase of community living, N!adima would not have done anything to enlighten him.

A standard set of anatomical terms are applied to all flowering plants:

> ≠*eidzi* (roots)
> //*oma* (bole, lower part of trunk or stem)
> *ji:/ado* (plant-trunk; "trunk" is also used for the human body, as in English)
> *ji:n//ama* (tree-arm, i.e., branch)
> //*o:* (bark)
> *g/inadzi* (leaves)
> //*a:* (flower)
> /*xudi* (seed)

All trees, shrubs, and most forbs have specific names as far as I could discover, but not all the grasses are named. Trees, shrubs, and the grasses of economic importance have nonderived names (Table 5). In many cases the name primarily signifies the part or organ that is utilized and its specific use is an extension of the meaning. This is similar to English usage, for example, "orange" primarily applies to the fruit and secondarily to the tree. In Glwi, as in English, fruit and tree can be distinguished by suffix, for example, *ka:* is the edible fruit of the gemsbok-cucumber (*Colocynthis naudinianus*) and the plant is distinguished as *ka:jisa* (gemsbok-cucumber-plant). When some part of the plant other than the utilized part is referred to, the name of the plant is prefixed to the anatomical term, for example, *n/oni//a:* (witgat flower).

By the time the Glwi I knew reached the age of puberty they were

79

Table 5. *Some plants utilized by the G/wi*

Scientific name	G/wi name	Portion used	Period available
Acacia giraffae	//ala	Root bark	Year round
		Gum	Year round
		Seed pods	April–June
Acacia mellifera	//k'uwa	Gum	Year round
		Branches	Year round
		Sticks	Year round
Acacia nilotica (=*benthamii*)	G!ō:sa	Root bark	Year round
		Gum	Year round
		Branches	Year round
Albizia anthelmintica	kxadu	Trunk	Year round
Aloe littoralis	Not known	Leaves	Year round
		Flowers	Winter
Aloe sp.	sa:sa	Stalk	Year round

80

Use	Method of use	Importance
Quivers	Straight section of root of suitable length and diameter is rapidly heated in fire to expand bark, which is then slipped off inner wood	Slight; *Ac. nilotica* much preferred
Trap bait[a]	Taken as exuded by tree	Moderate
Rattles	Two pods are struck together as accompaniment to singing and dancing	Slight
Sweet and trap bait[a]	Taken as exuded by tree	Slight; *Ac. nilotica* preferred
Firewood		One of many species used; smoke is slightly toxic
Lower firestick	Dry stick is peeled, trimmed, notched	Moderate
Quivers	Straight section of root of suitable length and diameter is rapidly heated in fire to expand bark, which is then slipped off inner wood	Great
Sweet and trap bait[a]	Taken as exuded by tree	Moderate
Firewood		One of many species used
Mortars	See Chapter 5, "Woodwork"	Moderate
Salad	Pounded up with leaves of *Terminalia sericea*, *Talinum arnotti*, and *Duvalia polita*	Slight; species is uncommon in game reserve
Delicacy	Flowers eaten, either straight from plant or included in salad with leaves of this and other species	Slight, although nectar and abundant pollen are likely to have nutritional significance
Stomachic	Dried and chewed when needed	Slight

Table 5. (*cont.*)

Scientific name	G/wi name	Portion used	Period available
Aloe zebrina	*g‖odu*	Leaves	Year round
		Flowers	Midwinter
Ammocharis coranica	Not known	Bulb	Year round
		Bulb juice	Aug.–Jan.
Antherpora pubescens	*pe:*	Culm	May–Aug.
Aristida ciliata	*g‖om*	Leaves	Year round
Asparagus sp.	*tsididi*	Leaves	Jan.–April
Bauhinia esculenta	*/xwi*	Seeds	May–Aug.
	g/am	Young tuber	May–Aug.
Bauhinia macrantha	*n≠an≠e*	Seeds	April–July
		Root	April–Aug.
Boscia albitrunca	*n/oni*	Fruit	Nov.–Dec.
		Trunk and branches	Year round

Use	Method of use	Importance
Salad	As *A. littoralis*	Moderate in dry seasons; common species
Delicacy	As *A. littoralis*	Slight; abundant nectar and pollen probably have nutritional significance
Antidote to small dose of arrow poison	Slivers of bulb are rubbed on site of wound; infusion of bulb is drunk	Said to be effective only against small dose
Coolant to inhibit water loss through perspiration	Bulb is shredded and juice is squeezed out over trunk and limbs to cool by evaporation	Used several times daily in hot, dry weather while resting
Main shaft of arrow	Stripped, trimmed, and straightened	Preferred material among five alternates
Tampon; postnatal umbilical dressing	Leaves are dried and crushed to soften them	Preferred of two alternates
Treatment of bodily pain	Leaves are boiled in water and applied as poultice to site of pain	Slight
To ward off approaching thunderstorm	Leaves are burned in cooking fire when storm is near	Slight
Food (starch, fat, protein)	Roasted, shelled; eaten as a dish on its own or ground and mixed with other foods	Great; occurs only in northern part of game reserve
Source of fluid	Eaten raw	Moderate
Food (starch, protein)	Roasted, shelled; may be eaten raw when still green	Great; staple when available
Fluid source	Chewed raw	Slight; holds little fluid
Food and drink	Eaten raw or agitated in mortar to separate juicy flesh from seeds to provide liquid	Great; staple when plentiful
Pestles, axe handles, knife handles	Adzed or carved	Moderate

Table 5. (cont.)

Scientific name	G/wi name	Portion used	Period available
Boscia albitrunca (cont.)	n/oni	Leaves	Jan.–Mar.
Brachystelma barberiae	//adu	Tuber	Jan.–July
Caralluma lutea	da:daba	Leaves and roots	Jan.–Aug.
Cassia (?italica)	kxei	Root	Year round
Catophractes alexanderii	gja:gjiba	Stick	Year round
Ceropegia sp.	g//aoma	Tuber	Jan.–July
Citrullus lanatus	n≠a:	Flesh and seeds of fruit	April–June; persisting to Sept. in a mild winter
		Juice	April–June; persisting to Sept. in a mild winter
Citrullus naudinianus	ka:	Flesh of fruit	Feb.–May
Coccinia rehmanni and C. sessifolia	G/a:	Tuber	Year round, but mostly winter to Nov.
Combretum imberbe	kxedu	Gum	Year round
		Wood	Year round

84

Use	Method of use	Importance
Relieve dryness of mouth	Chewed raw	Slight
Food, fluid	Eaten raw or roasted	Greatly favored but not plentiful
Salad	Eaten raw, pounded up with leaves of *Terminalia sericea* and *Talinum arnotti*	Slight; common only in eastern part of game reserve
Treatment of liver complaints	Root scraped down to yellow stem; boiled and drunk as hot infusion	Slight
Upper firestick	Straight stick is peeled and trimmed	Moderate
Food, fluid	Eaten raw or roasted	Greatly favored but not plentiful
Food (rich in vitamin C; seeds rich in protein) and fluid	Flesh pulped; eaten raw or added to other foods, including meat, to make stews; seeds roasted and ground to meal. Juice also used in preparing hides, which are dried; process repeated 3 or 4 times. Juice also used to wash hands before eating or after working with arrow poison	Very great; main food when plentiful
Food and fluid	Fruit roasted and peeled; flesh pulped. Eaten alone or with *Grewia* berries	Great
Food (starch) and fluid	Roasted and chewed; fibers are spat out. Tuber is poisonous when raw. Unripe fruits of *C. sessifolia* occasionally eaten raw or boiled	Very great; staple in dry season
Sweet	Eaten raw	Slight
Slow burning firewood		Moderate; species has limited distribution

85

Table 5. *(cont.)*

Scientific name	G/wi name	Portion used	Period available
Commiphora pyra-canthoides	/udi	Root	Year round but more often winter and early summer
Crinum sp.	//ou	Bulb	Year round
Cucumis hookeri	n≠o'nu	Flesh of fruit	April–July
Cucumis metuliferus	Not known	Flesh of fruit	April–July
Cyphia stenopetala	Not known	Tuber	April–July
Dipcadi sp.	n!om	Bulb	Feb.–July
Duvalia polita	//aija	Leaves	Jan.–July
Ehretia rigida	g//a:	Fruit	Jan.–Feb.
Elephantorrhiza elephantina	//adidi	Root	Year round
Grewia avellana	Not known	Fruit	Jan.–June
G. flava	kxwam		(seasons of
G. flavescens	kxwam		species over-
G. occidentalis	kxaokxwam		lap slightly)
G. retinervis	//ani		
Harpagophytum procumbens	Not known	Root	Autumn, winter
Hibiscus irritans	≠xade	Root	Year round
Hydnora sp.	kaigu	Whole	Late winter

86

Use	Method of use	Importance
Fluid	Root is split and juice is sucked out of pith. Singularly refreshing, but much work entailed in digging up root	Moderate, because of patchy distribution of shrubs yielding appreciable amount of fluid
Juice of bulb as body coolant and for softening animal hides during tanning	As *Ammocharis coranica*	Moderate
Fluid	Raw or roasted, peeled	Moderate; there are several more pleasant-tasting alternatives
Food and fluid	As *Citrullus naudinianus*	Slight; common only in northern dunelands
Fluid	Roasted	Slight; tuber is small
Food and fluid	Roasted	Slight
Food and fluid	Salad, pounded up with leaves of *Terminalia sericea* and *Talinum arnotti*	Moderate
Food	Eaten raw	Slight
Dye for tanned leather	Root shavings are rubbed into hide while moist	Moderate
Food	Raw; fruits are sucked and seeds are spat out, or fruits are pounded up with other foods	Great; staples in their seasons
Fluid	Chewed	Slight to moderate
To induce cosmetic weight gain (apparently ineffectual)	Dried and chewed	Part of every woman's toilette
Food	Eaten raw	Slight; uncommon in game reserve

Table 5. (cont.)

Scientific name	Gwi/name	Portion used	Period available
Ipomoea trans-vaalensis	Not known	Tuber	Jan.–Sept.
Kalanchoe sp.	g//am//am	Root	Year round
Lapeirousia	n//u:	Corm	Mar.–July
Lasiophon kraussianus	!adibe	Root	Year round
Leonotis microphylla	//udi	Stalk	April–Sept.
		Leaves	April–Sept.
Mormodica balsamina	Not known	Flesh of fruit	Jan.–Mar.
Nerine sp.	Not known	Bulb	Aug.–Jan.
Ochna pulchra	g!hala (alias !ara)	Fruits	Dec.–Jan.
		Trunk	Year round
Raphionacme burkei	bi:	Tuber	Year round, but mainly July–Nov.
Rhigozum brevi-spinosum	ta:	Straight branches	Year round

Use	Method of use	Importance
Food and fluid	Eaten raw or roasted	Moderate; species not very common
Emetic for relief of stomach pain	Root peeled, boiled in water; infusion drunk	Slight
Food	Roasted	Slight; species not common in game reserve
Treatment of body pains	Roots are burned; ash is powdered and rubbed into cicatrices incised at site of pain	Slight to moderate
Stalk: as a tube to suck water from hollow trees	Dried; pithy core is removed, leaving a hollow tube	Slight
Leaves: as occasional tobacco substitute	Dried and crushed	Slight
Food	Peeled, eaten raw	Slight; eaten only as occasional snack
Coolant	As *Ammocharis coranica*	Moderate
Food, oil (see *Ximenia caffra*)	Fruit eaten raw; pips boiled in water to extract oil	Slight to moderate
Mortars and bowls		Moderate; occurs mainly in eastern part of G/wi country
Fluid source; ritual washing of infants and of husband and wife at latter's menarchial ceremony	Tuber is shredded, pulp is squeezed and drunk; pulp is also used as washrag	Great; staple in dry season
Digging sticks	Peeled, trimmed, and sharpened	Great

Table 5. (cont.)

Scientic name	G/wi name	Portion used	Period available
Sansevieria scabrifolia	*g/wi*	Leaves, bulb	Dec.–Oct.
Scilla sp.	*dzu*	Bulb	Jan.–July
Strychnos cocculoides	*g/ua*	Fruit	Jan.–July
Talinum arnotti	*//abe*	Leaves	Jan.–April
Talinum sp.	*n≠ao*	Leaves	Feb.–April
Terfezia pfeilli	*kutsi*	Whole	Jan. or April–May
Terminalia sericea	*g≠a:*	Leaves	Nov.–Jan.
		Straight branches	Year round

Use	Method of use	Importance
Leaves: string; bulb: fluid	Leaves decorticated by scraping sharp edge of split stick down length of leaf; retted fibers are spun into string while still moist. Bulb is chewed raw for fluid	Great, for fiber; slight for fluid
Food and fluid	Roasted, peeled, and eaten	Great
Food and fluid	Eaten raw	Moderate; occurs mainly in eastern part of G/wi country
Food and fluid	Eaten raw, alone or with leaves of other species	Moderate
Condiment (salt substitute); remedy for toothache and dryness of mouth and throat; an excellent remedy for insect stings	Leaves chewed; crushed leaves are applied to site of insect sting	Moderate
Food, some fluid	Roasted	Great when plentiful
Food and fluid	Young leaves eaten raw, usually pounded up with leaves of other species	Moderate
Medicinal	Pounded leaves applied as poultice to wounds, ulcers, or site of pain; leaves chewed as stomachic	Moderate
Springhare sticks	Peeled, trimmed, spliced together to form 4 to 5 meter rod	Great

Table 5. (*cont.*)

Scientific name	G/wi name	Portion used	Period available
Terminalia sericea (*cont.*)	*g≠a:*	Straight branches (cont.)	
		Bark	Nov.–June
Tribulus terristris	Not known	Leaves	Jan.–Mar.
Trochomeria macrocarpa	Not known	Fruit	Jan.–Feb.
Turbina oblongata	Not known	Leaves	Feb.–May
Vigna dinteri	*om/i*	Root	May–Aug.
Vigna sp. (*?triloba*)	*kwam/i*	Root	May–Aug.
Walleria nutans	Not known	Bulb	Feb.–June
Ximenia caffra	*≠odi*	Fruit	Jan.–Mar.
		Seeds	Jan.–Mar.
Zizyphus mucronata	*≠kxado*	Branches	Year round
		Fruit	May–June

ª See also text, p. 215.

92

Use	Method of use	Importance
Link shafts of arrows	Peeled, trimmed, carved to requirement	Moderate; substitute for bone
Pegs for drying hides during preparation	Peeled, trimmed, carved to requirement	Great
Bindings	Lengths of bark peeled off wood, used while moist	Moderate
Glue	Chewed; juice is spat out and applied directly	Moderate
Tanning material	Infusion from bark applied to hide	Moderate
Laxative	Chewed and juice is swallowed or decoction is drunk	Slight
Food	Added to meat stew	Slight
Food	Eaten raw	Slight
Additive to snuff	Leaves dried, ground, added to powdered tobacco	Slight
Food and fluid	Eaten raw or roasted	Great; highly favored
Food and fluid	Roasted	Moderate
Food and fluid	Eaten raw or roasted	Slight
Food and fluid	Eaten raw	Moderate
Oil for skin comfort, for treating wounds and ulcers, for preserving leather and wood	Oil extracted by roasting and crushing seeds between hard surfaces	Great
Axe handles	Suitably shaped portions peeled and trimmed	Slight
Food	Eaten raw	Slight

able to identify and name all local trees and shrubs and most named herbs. Specimens of leaves, fruits, tubers, and flowers were usually readily recognized. Story (1958, 1964) has remarked on the precision of Bushman botanical taxonomy, including that of the G/wi, at the species level and of the paucity of groupings at higher levels. In my observation, the specific taxonomy of the G/wi is, in some instances, more discriminating than that of European botanists (e.g., both *Ximenia caffra* and *Colocynthis naudinianus* are split into two species each, according to whether the fruits are sweet or bitter; identification is made on visual examination of external characteristics, not on taste). Even though there is no terminological recognition of genera, some informants spontaneously described similarities that exist between closely related species within a genus (e.g., some of the acacias) and commonly resorted to the model of the G/wi kinship system to explain the similarities. I became aware of only two categorizations, *gera:dzi* (esculent plants), the residual category being unnamed, and *g/a:dzi* (grasses).

The G/wi also have a good knowledge of the composition of plant communities and use it to narrow the search for individual species. Information was frequently volunteered that a particular species was typically found in association with other named species or at particular sites classified by soil type, topographical characteristics, and so on.

A scale of preferences exists among esculent plants, which form the subsistence base of the G/wi. This scale is of great economic importance and is the most signficant single factor governing migrations. However, excepting the ripening of a prolific growth of tsama melons, plants have none of the emotive significance of game animals but are discussed and dealt with as a routine necessity of life. The gathering of plants requires much knowledge, which is shared by men, women, and children without esotery, but involves little skill once the specimens are located. Because the immobility of plants reduces the number of variables involved in collection, a gatherer needs to know their location and their season of availability and to have the ability to recognize useful species. The rest is a matter of hard work, and apart form this, the yield of the season is in the hands of N!adima, who cannot be influenced to make it more plentiful. Man's role is that of exploiter only. These aspects, and the fact that plant foods are not exchanged as gifts and thus have less social significance than meat, perhaps explain why food plants are regarded with general emotional neutrality.

The spatial ambient

The three-tiered structure of the universe has been mentioned: the land, the sky country, and the underworld. Man's doings and the extent to which his actions have rationally comprehensible consequences are confined to the land, the effective compass of his knowledge. It is also the region in which are located the comprehensible forces that affect the lives of the G!wi, the action and interaction of life-forms, weather, heavenly bodies, and the fabric of the land. Knowledge of the underworld and sky country is fragmentary and uncertain, and although interregion interaction occurs when the subterranean monsters are angered or when N!adima is moved to intervene, the upper and lower regions are beyond the reach of human means and are peripheral to the G!wi cognition of, and interaction with, their environment.

The land

The full extent of the land, $n \neq o{:}ma$ (world, country, territory, district), is unknown. Few G!wi have any knowledge of geography beyond a radius of 250 km and the personal experience of most is limited to a range of about 80 km. They believe that they live in the middle of the inhabited world between the "far countries" in which the sun rises and sets (because, as they point out, it reaches its zenith over their own country halfway in its daily journey). The world, they believe, stretches in all directions to its finite but unknown boundaries, and there is little speculation about its nature beyond the parts known to them. Assuming the remainder to resemble their own country, the G!wi make sporadic imaginative constructs concerning it when a perceived intrusion demands some explanation. For instance, the ≠xade people correctly deduced that I was not a native of Ghanzi, to which I regularly returned for provisions, but must come from the southeast. The deduction was based on my use of aircraft; airplanes that overflew ≠xade in the morning were usually seen to return on a southeastward heading in the afternoon – "they were returning home to roost." As I used aircraft, I must come from the country in which they "roosted." I am somewhat larger and stouter than any G!wi. To them, my size meant that my country must have been more fruitful than theirs, a conclusion supported by the fact that their rains come from the northeast and any country to the east must therefore be better watered than their own.

95

Sand covers most of the central Kalahari and consequently comprises the main part of G/wi surface and immediately subsurface geology. *Xwamsa* is the generic term for sand, soil, and earth. *G≠o:* (clay or tufa in the dry state), *//asa* (wet clays and tufas), and *!'kade* (decayed calcrete) are further distinguished and red and white sands are identified by use of the generic term with adjectival qualification, *!'u:xwamsa* (white sand) and *n/uwaxwamsa* (red sand).

The only solid rocks found in the central Kalahari are the quartzites of the Ghanzi beds (which protrude through the sand mantle in the northwestern corner of the reserve to form the Tsau or Sonop Hills) and the extensive strata of gastropod calcrete and diatomite found in nearly all drainages and pans. There is no generic term for stone, the two types of which are called *n//wasa* (quartzite) and *!ado* (calcrete, diatomite).

The sole G/wi essay in geomorphology is a mythical account of the genesis of the Okwa valley. G//amama was bitten in the leg by a python near Gobabis, far to the west. The bite was very painful and he became ill and feverish and very thirsty, so he went in search of water. He headed eastward in the direction of the Botete River. As he walked, G//amama dragged his injured leg, gouging out the present valley. The fever made him nauseous and from time to time he vomited. His dried vomit formed the calcrete- and diatomite-floored tributary valleys of the Okwa (which itself has a sand bed). The twists and turns of the main valley were occasioned by the misdirection and harassment of G//amama by a variety of animals that took this opportunity to avenge some of the wicked things G//amama had done to them (see Silberbauer, 1965:96). This story includes country far beyond G/wi ken, even beyond that of their western neighbors, the !xõ:, suggesting the foreign addition of detail. However, it is curious that G//amama headed east in search of water, as Gobabis, where the python bit him, is on a bend of the Nossob River, the site of a widely known perennial spring. This fact indicates an essentially G/wi origin to the myth, as they do not know about the water at Gobabis.

The earth is regarded as the inert, unchanging matrix in and upon which life exists. There is no comparable account of the formation of the other major valleys (the Passarge Pans chain and the Deception and Merran valleys): These and other features were fashioned by N!adima at the time of creation and, with the rest of the fabric of the universe, have remained unaltered since then (excepting that which G//amama wrought).

The known world consists of the territory of an individual's own band, territories of allied bands and those that he has visited, plus the areas of which he has secondhand knowledge gained from acquaintances who have traveled farther than usual on trading expeditions. Interband migrants and visitors bring the information up to date and sometimes can tell of territories unknown to their hosts. Even across a radical language boundary (e.g., between G/wi and !xõ:) there are bilinguals who move between communities and pass news on across the language boundaries. Known and unknown parts thus merge into a continuum of knowledge and interest, which decline with distance from the individual's home.

The flat, rather monotonous desert topography does not have many prominent features that could readily be classified. *K'ajewama* is a general term for the countryside around and beyond human habitation, corresponding to the South African term "veld." Vegetational cover is a criterion of classification of topographical types: */xaosa* is high thorn woodland of the type found in the northern dunelands of the reserve and flanking the Okwa valley east of Easter Pan; *g/wisa* is low, mixed woodland – what is called bushveld in South Africa; and *g/a:dzi* is grass and scrub plain. The term */amsa* (wilderness, thirstland) is applied to any locality badly affected by drought and also to uninhabited country with sparse vegetation.

The main surface features are:

djuausa (relatively narrow, steep-sided valley)

/xa:sa (relatively broad, shallow valley)

gjiusa (path, drainage line connecting pans, a narrow valley; latterly, a road made by repeated passage of vehicles)

g≠eima (extensive exposed stratum of gastropod calcrete; in the reserve these are confined to pans and drainage lines)

djeusa (a natural or artificial pit, usually in diatomite outcrops)

tshasa (literally, water; a depression on a pan into which rainwater drains)

!o'osa (pan; a shallow, hard-floored depression)

Excepting the exposed strata of calcrete and diatomite and the Sonop Hills, the reserve is covered by a uniform mantle of sand, the monotony of which is broken by pans, drainages, and valleys. Each of these is named and provides a geographical reference point. The names are varied, and their origins and, in some cases, primary meanings have been lost. What I have called Easter Pan is rightly called //a: (bat-eared fox), and there is Tsho:khudu: (hot hand), Tsxobe (to pick up something from the ground), and G!õ:sa (*Acacia benthamii*; quiver), but nobody could tell me the origin of these names or the literal meaning of Kxaotwe, another pan.

97

These features are too few to provide an adequate number of reference points and are supplemented by large trees and thickets and other features of a durable and, to the trained eye, distinctive nature. These are given names as diverse in meaning as those of pans. Localities are synonymous with the features, which are seldom more than 5 km apart, many being separated by only about 2 km. They form an irregular grid across the whole of a band's territory and are the means whereby any position can be fixed with a fair measure of exactness and without ambiguity. Although the system baffles the stranger, it is efficient for those who know and use the country, that is, members and visitors of the band occupying it.

Direction is not expressed in absolute terms (e.g., north, east, etc.) but in terms relative to named localities. The choice of orienting locality varies with the geographical scale. To exemplify: Within a few miles of ≠xade Pan, northeast is termed *g//akokum n!u* (the direction of G//akokum, a string of pans some 19 km distant). When greater distance is involved, the referent is Tsxobe, a tributary valley about 90 km to the northeast. For yet greater distances, for example, when referring to the direction of the origin of summer rains or the direction in which game herds migrate after autumn, the Botete River is the place of orientation. The directional meaning of the referent alters, of course, with changes in one's own position. Thus, although *g//akokum n!u* means northeast when at ≠xade Pan, it means due south when at Piper Pans. Confusion is obviated by the use of referents appropriate to the scale.

Distance is measured by the time taken to walk it. As speed varies according to circumstances, knowledge of these must be implicit in the measurement. A woman collecting food may cover 1 or 2 km in an hour, but a man setting out for a hunt could cover 6 or 7 km in that time. Under the ideal conditions of autumn, with good reason for pressing on, adults can walk 60 km in a day, but a household, without reason for haste, averages a bit more than 4 km per hour and walks for some eight hours during the day, that is, covers less than 40 km per day. The latter is the most frequently used standard when measuring long-distance travel, as these are the circumstances under which it is normally undertaken. The total distance is expressed as the number of nights one would sleep while making the journey, and fractions of a remaining day are indicated by pointing to the position in the sky of the sun on the arrival of the traveler.

The optimal route is not a straight line but a compromise between distance, ease of travel (e.g., avoiding thick bush but taking advantage of shade), and the convenience of en route food supplies. Most

guides could readily accommodate to the quite different requirements of vehicles, including the geometrical problems of the shortest
straight-line route from an established track to a point some distance
from it.

Among my informants were men who had very good directional
sense; after following a roundabout, looping course from ≠xade to
Ghanzi, which the informant had not previously visited, I asked him
to point out the tree on the horizon under which he would pass if
heading straight back (note: not a walker's most suitable route) and
he was correct to within half a degree of arc. Other informants, when
flying for the first time, remained perfectly oriented and afterward
could accurately recount the track flown in terms of known localities.
(In twenty years of transporting passengers, I found this to be an un-
common ability.) They were also able to relocate themselves on returning to known country after flying over strange territory (another
very rare ability). It is clear that the methods used by the G/wi provide them with a frame of geographical reference in which space and
position are conceptualized with a high degree of exactness and accuracy.

Attitudes regarding territory are affected by the fact of N!adima's
ownership of the land and its resources and by the fact that the primary bond is between the individual and his band, whereas the link
between the individual and territory is derived from the bond between community and land. The secondary nature of this link is consistent with, and perhaps consequent upon, their being N!adima's
sufferance tenants who apportion the share of land and its resources
among themselves. They have no higher legitimizing authority than
consensus among those involved and no means of resolving conflict
other than by coming and adhering to common agreement. Rights to
a territory flow from band membership and can be exercised only
while acting as a member; to leave the band is to relinquish rights to
its territorial resources. Visitor's privileges to use resources, although formally extended by the "owner" of the territory (see Chapter 4), require the previous agreement of the band. A girl, toward the
end of her menarchial ceremony, is introduced to the territory, its
fruits and other resources by her band fellows. This I interpret as
being symbolic of the primacy of the social bond over the link between individual and land. Rights are created when a pioneer nucleus moves into a new territory and are extinguished when a band
abandons its country (e.g., in times of disaster), with no residual
rights remaining with the survivors who migrate to other bands.
(This must be distinguished from the preferential membership privi-

lege extended to one's parents, other close kin, and friends.) There is manifest fondness for the home territory and people are reluctant to be away from it for any length of time. However, this reluctance is not evident in those migrating to join another band, which also argues for the primacy of the social bond.

Although there are perceived dangers and hardships, the foremost being the sun and its heat, the land is regarded as being hospitable. The G/wi know full well that early summer is a lean season in which food is scarce and existence is uncomfortable, but they expect the land to furnish the means of survival and regard late summer and autumn, when the climate is equable and food is plentiful, as the norm – the time when man and land are in their proper relationship. This is the reference point and other seasons represent a departure from it.

The troposphere

The G/wi include within the middle, or earthly, region of the universe the space between the ground and the sun, moon, and stars. The heavenly bodies are believed to be situated just above the highest clouds. This zone coincides with the troposphere. It contains the weather systems and the sun, phenomena of obvious importance that constantly come up in G/wi conversation. In combination they result in another recurring topic, the seasons. The seasons are not regarded as independent entities or even as sharply demarcated aspects of one entity. Seasons are an aspect of the land and consist of aggregations of factors: the state of vegetation and the availability of food, temperature, wind, presence or absence of water, the behavior of animals and of men. But their qualities are most clearly seen in the weather characteristic of each, and for this reason, weather and the seasons are discussed as one topic under the broad heading of the land.

Weather and the seasons

Weather is seen as a systematic phenomenon, a unified whole composed of parts whose workings can be seen and are, up to a point, explicable and predictable. The prime cause is N!adima's will, manifest in the creation of the weather system, but he is bound by his own ordinance in that he can manipulate the weather system or its parts only within the normal range of behavior. (As the weather of the central Kalahari is violent, erratic, and greatly varied, he has allowed himself generous scope for manipulation.)

The normal precursor of a wet season is the strong northeasterly wind that blows throughout spring and early summer. Although this hot, dry wind brings great discomfort to the G/wi, they endure its blast with fortitude, sustained by the belief that it "sweeps the rain together and heaps up the clouds." The longer and harder it blows, the more rain will be swept up to fall in the coming second half of summer. When the wind has done as much as it can, it tires and ceases to blow. Released from this pressure, the rain-bearing clouds rush back across the desert, coming out of the southwest and dropping the first, light showers of summer. From the northeast they move again across the desert at a more leisurely pace, growing fatter and fatter, and drop the normal thundershowers of the proper wet season.

The systematic nature of the conceptualization of weather is also indicated in the short-range forecasts, which are based on the occurrence of conditions believed to be indicative of certain types of weather. These conditions do not have causal roles but are prodromal, being parts of what is to come. The south wind, for instance, which heralds cold snaps in winter, does not bring the cold but is "the wind of the cold," its vanguard and an integral part of the whole phenomenon. The humidity, which lends unmistakable freshness to the morning air, and the wind and close heat of the forenoon do not cause afternoon thunderstorms but are "the smell, the wind and the heat of the rain."

A thunderstorm is spoken of as being a leopard; thunder is its growl and the sound of crunching bones as it eats, and lightning is the flashing of its eyes. In European meteorological theory, thunderstorms occurring in unsaturated air masses tend to have highly variable tracks (Donn, 1951:83 ff.; Air Ministry, 1960:102 ff.). In G/wi theory, the leopard possesses a measure of volition and its current mood is revealed in the pattern of its movements and is forecast by interpreting the precursory signs of storms. When it is calm, the storms track straight, but when it is angry or flustered, its movements are erratic. A sudden and, to me, quite unexpected deviation of a storm, which, until sunset, had headed steadily toward the camp, caused the band members little surprise. It had intended coming straight, they said, but was in a bad mood and did not pay attention and got lost when darkness fell. On another similar occasion, the leopard, thought to have been diverted by an antelope that it had failed to catch, had chased it all over the countryside. (Although perhaps not orthodox meteorology, it was certainly a graphic way of describing the way in which the towering mass of cumulonimbus charged about the evening sky.)

An appreciation of causation and the interrelationship of meteorological phenomena was clearly indicated by the explanation of a sudden shift of wind and sharp drop in temperature one steamy summer afternoon. A thunderstorm was in progress some 16 or 20 km to the north, and although it is common to experience a chilly first gust in the path of an approaching storm, it is unusual for this to happen when located at right angles to its track. Informants pointed to the dark and light sheets of falling rain and hail and explained how so much coming down must disturb what is beneath it, "the rain pushes the air aside, and the hail cools it; when there is so much the air cannot all go in front and some must go out at the side like this."

This explanation is not far removed from the European view of an accelerating downdraft in a mature cumulonimbus cell being cooled below the temperature of its surroundings so that the outflow sets up a miniature cold front, giving an increase of wind speed and a change of direction, which is influenced by the distribution of surrounding mature cells (Air Ministry, 1960:103–109). A meteorologist would not accept the identification of storm and leopard, but the inconsistency is perhaps more apparent than real. The G/wi account is based on observation of a limited range of phenomena within a narrowly circumscribed geographical area, and the observations are qualitative rather than quali-quantitative. The G/wi have attempted to explain what they have observed in their environment and have succeeded in formulating sets of causes and sets of effects that result in predictions of fair accuracy. They have not proceeded to primary causes, but it is evident that they consider their environment or, at least, aspects of it to be capable of rational explanation.

All life in the Kalahari depends on rain, a fact that the G/wi keenly appreciate. They have an intense preoccupation with rain, but they have not, however, made any attempt to influence it, for example, by magic or prayer. Their role is passive, for, ultimately, the weather is governed by N!adima and they can only try to avoid angering and provoking him into keeping the rain away from them. It is true that an imminent storm is addressed and asked to come soon and to drop its rain not only in one territory but everywhere, so that people everywhere may have food; "if you fall on an ugly woman, you will make even her beautiful." These praises and entreaties are not believed to encourage the rain but are intended to avoid affronting it and N!adama by a seeming lack of appreciation. Other praises are less directly flattering but are nevertheless indicative of appreciation and hopeful anticipation:

Dju:ma am kho e g!udi, n/im/amka hi ≠u:
Ixansi hi ≠u:, n/im/am be ≠u:
(Rain his skin it is many, this-day-on [it] will rain
excessively will rain, this day he rains.
[or] Cumulus clouds are many, it will rain today;
it will pour with rain this day.)

Informants rejected the suggestion that such praises influenced either the rain or N!adima and emphasized that it was necessary to take pains to avoid being thought of as unappreciative by N!adima. His ability to control rain was demonstrated in 1974. The rains had started early in October 1973 and were inordinately heavy and prolonged. Not only were great inconvenience and discomfort caused by the ensuing floods, but a plague of *Anopheles* mosquitoes bred in the enormous stretches of water that lay everywhere for months and months and brought a serious epidemic of malaria in which many people died. G/wi on the northeastern ranches told me that they had discussed the problem of the floods among themselves and with other Bushmen and had decided that the best thing to do was to kill a chameleon (an act that normally angers N!adima and provokes him to "dry up the land"). They were certain that N!adima would become very angry if an adult did the killing but would understand if a less-informed, younger person performed the act. The older men persuaded two boys to kill a chameleon, and to make it clear to N!adima that the act was a considered and deliberate one, and not thoughtless, casual, or accidental, the boys were made to slit the chameleon open and place it on a board, belly upward and limbs outspread, out in the open where it could be easily seen – a sort of ad hoc ritualization. The rains duly ceased. For the remaining weeks of my stay the men were anxious about the eventual outcome – would N!adima be angry? Would they and/or the boys be punished? Would there be rain the following summer? This uncertainty was consistent with their statements that this was an innovation devised in desperation. The affair seemed to confirm my earlier interpretation that man is expected to work out his own means of survival and that, in limited fashion, rational experimentation leading to novel solutions is legitimate.

The central concept in rainfall is *dju:* (rain) and not *tshasa* (water). Rainwater is only termed *tshasa* after it has fallen and has gathered in pools. It is *dju:* that brings life to vegetation and to the earth generally, and all the actions of a storm are performed by *dju:*, which ≠*u:* (rains), *kjibi* (flashes lightning), and *kjidi* (thunders). That which is cold and chills the wind or wets that on which it falls is *dju:*. When I experimented with the meaning of the word and the extent of the concept and said that I would drink *dju:*, the memory of the remark awoke laughter for weeks afterward. They explained to me that in order to achieve this feat I would have to swallow the entire complex of clouds forming in the approaching zone of storms, indicating that the concept embraces all phenomena directly associated with precipitation. Two types of rain are identified: *dju:ma* (masculine) is the hard, driving rain that falls from thunderstorms and "speaks loudly"; *dju:sa* (feminine), is the steady, soft, set-in rain that sometimes covers the whole central Kalahari – "this is good rain, which speaks softly and is gentle."

The other phenomena of storms are of less importance. Although there are wonderfully impressive displays of lightning and many trees are struck during a season, human casualties are rare and no fear is shown of either lightning or thunder (it is thunder that is believed to wreak damage). Hail presents a slight danger when large stones fall. Superficial injuries are sometimes inflicted, and I was told of one person who was killed by hail.

Two cloud types are terminologically distinguished, *dju:khodzi* (literally, rain-skins), clouds of the cumulus family, and *dju:lodzi* (literally, rain feathers), clouds of the cirrus family. Both are moved by wind (cf. the description above), but when the former clouds have developed into mature storm cells, they have a measure of volition and can manipulate wind, as described above. Only one wind had been named, that is, *nllabem* ≠*a:ka* (giraffe wind), a cold easterly, which is often accompanied by set-in rain in summer. Other unnamed winds have male gender if accompanied by warm or warm and wet weather and are regarded as good and life-bringing. Cold or cold and wet winds (excepting the easterly) have feminine gender and are regarded as dangerous or at least as being sterile. High temperatures are caused by the sun, but low temperatures are associated with the feminine winds, which are believed to be part of the whole cold-weather complex. Both extremes of climate are disliked. Hot weather is more common and stressful, and complaints against it are therefore more frequent, but my impression is that cold weather is equally unpopular.

In counting the seasons, the G/wi do not have any particular starting point, but the passage of years is reckoned in terms of the number of winters, which may therefore be regarded as the end of the

G/wi year, and in this description of the seasons, the season following winter is taken as the beginning of the G/wi year.

> !hosa, the name is specific and signifies this season. "It is the hot
> time, the time when the trees flower" (i.e., *Acacia mellifera, Ac. ne-
> brownii,* and *Boscia albitrunca*). This is the dry season between the
> end of winter in August or September and the first rains in late No-
> vember or December, that is, it is defined in terms of climatic and
> vegetational phenomena and not as a fixed period of time measured,
> for instance, by reference to stars, although the stellar association is
> recognized.
> N/laosa, "the rain time when the grass is green and the antelope
> breed" (sic; many drop their young in *!hosa*). This season runs from
> the onset of rain in late November or December until late March or
> even into the beginning of April.
> Badasa, "when the tsama melons are plentiful," from late March or
> early April, when the rains cease, until the veld begins to dry out
> and the nights become cool, which is about the middle of May.
> G!wabasa, "when the plants begin to die." The ephemerals die and
> the deciduous species drop their leaves. This season lasts from
> about mid-May to mid- or late June.
> Saosa, "the cold time," from mid- or late June until about the end of
> August.

In some years there is a fair fall of early rain in the beginning of
October, which is usually followed by a dry, hot spell, after which
the rains become properly established. When this happens, the early
rains and their aftermath are called *badasa* (i.e., the same name as the
early autumn season). This may last until early January but is usually
over by December. Then follows *n/laosa,* and the succeeding season
is called *g/lobadasa* (great badasa).

Attitudes toward the seasons are directly related to material well-
being and physical comfort. Because of its heat and the scarcity of
food, *!hosa* is the most difficult to bear. People become morose, depressed, and listless, and it is a time of disease, discomfort, and dis-
interest. The season's burden is aggravated by the lack of the company of band fellows, as the band is nearly always in the separation
phase at this time. The lack of company is sorely felt, as the principal
pleasures in life derive from, or are dependent upon, the presence of
friends and kin; consequently, this season and its hot sun are bitterly
resented.

Children feel the isolation of this season as keenly as do the adults.
They become bored with the inactivity enforced by the sun's heat
during most of the day, miss their friends and the games they play
together, and become fretful. Children are less resistant to the mental

and physiological strains than are the adults, whose experience of many seasons has better equipped them mentally and has taught them to accept the conditions with a measure of stoicism, whereas the children have not yet acquired that self-discipline. The greater physiological stress in children is occasioned by the denial of their relatively more critical nutritional requirements and their more nearly critical fluid imbalance, which is caused by perspiration (children's higher skin-surface to body-mass ratio causes them to perspire at relatively higher rates than do adults). The demands that children make on their parents at this time are very heavy, but the burden is borne with great patience.

The only relief afforded is the anticipation of the coming rains, and if *!hosa* goes on for longer than normal, people become anxious lest the rain come too late and be insufficient for a good growth of food plants. This could mean complete famine and death. I have not seen a serious threat of famine, but in years when conditions were very poor, although anxiety was obvious, there was no indication of panic or of hope abandoned. There seemed to be durable faith in the ability of their environment to sustain them at survival level at least, although not always in comfort.

N//aosa brings immediate relief from anxiety, although hunger and thirst continue for some time as plants must first recover before man and animal can utilize them. As the season becomes established, the band re-forms, and added to the joy of having rain, there are the pleasures of company. The variety and duration of activities increase with the greater amount of food available and with the amelioration of the sun's heat by clouds and emerging foliage. The lassitude of the earlier season disappears and people show an increasing interest in their now more rewarding environment, as their emotional state lifts steadily with the improving material circumstances.

Badasa is the peak season in which the Kalahari is at its most fruitful and the weather at its best for man. This, the time of relative plenty, is the good season, the "right" time of the year when, in the eyes of the G/wi, the environment is as it should be, other seasons being in varying degrees detractions from this state and deprivations of what the environment should hold for man. This seems a remarkably cheerful and optimistic attitude, as *badasa* is a short season of only six to eight weeks. In this season people speak warmly and spontaneously of their country and how good it is.

Because of the amount of food available and the clemency of the weather, travel is easy in *badasa*. The land can support a greater density of population than in other seasons and there is much visiting between bands. The welcome company of visitors adds to the pleasures of plentiful food. Less time and labor are needed to collect a day's ration, and with the cooler weather, food lasts longer before becoming desiccated or rotten. It is possible to collect two days' supply of food in one sortie, leaving the next day free for games, dancing, and conversation. In very good years, this season takes on some of the character of a marathon party, with the people in appropriate mood.

As autumn begins to fade and nights change from cool to chilly, *badasa* is followed by *g!wabasa*. Initially, the people are still cheerful and there is much boisterous happiness, but they gradually become quieter and more serious as the approach of winter is more clearly borne on them. They calmly accept what must come – the freezing nights of winter and the hot, hungry days of early summer – confident that the land will yield enough to enable them to survive until the rains return.

Cold weather (*saosa*) has no redeeming features in the eyes of the G!wi, and they are miserably uncomfortable in the subzero temperatures of winter nights. The days are pleasantly mild when the south wind is not blowing and the season is borne cheerfully enough. Food is less plentiful but is seldom so scarce as to impose real hardship, and *saosa* does not engender the depression and resentment, amounting almost to hatred of the discomfort their environment imposes, that is seen in *!hosa*.

The firmament

The sun is the only heavenly body that is of major importance to the G!wi. They focus on it much of their resentment of the miseries of *!hosa*, and complaints that "the sun is killing us" or "the sun will kill us" are frequently heard. "We fear the sun" might almost be the motto of the G!wi, so often is the phrase heard as a complaint and as a reason for not doing things during all but the winter months. The sun is the cause of any misfortune arising from drought and heat, which themselves are brought about by the sun.

The word *!amsa* has the significance of "sun" and of "day," both as a unit of time and as contrasted with night, and also means a hot, dry inhospitable region, a drought-stricken wilderness.

The sun has no capacity of autonomous action but is controlled by N!adima. Although the G!wi address a short formula to the sun, "*A dzenene dza:, a dzenene dza:, n!lam, e !ho: 'e*" (It is terribly hot, it is terribly hot, pass on, this is *!hosa*), they deny that this is a prayer intended to persuade the sun to hurry on and bring the next season. As was the case with the formulas said to the rain, this one is to show respect for so powerful a creation of N!adima's and to make it clear that the sun is not being slighted by the complaints about the hardship that it causes. Consistent with this statement is the fact that it is regarded as a serious matter to curse the sun. One may complain about the sun and the *"dzenene dza:"* formula will set matters right, but the idea of uttering stronger imprecations caused acute concern; such an action would anger N!adima and he would do something far worse than that which the summer sun normally does.

When the sun sinks in the west, darkness is caused by the shadows of the trees of that far country. "N!adima makes the sun come down to that country so that it may cause these shadows and there may be darkness for man and animal to sleep and for the nocturnal animals to move about in. N!adima does this also so that he may catch the sun and eat its top [or outer] body." It is not clear whether he does this to derive nutriment or whether he does it to prevent the sun from getting up again in the night. Both reasons were advanced separately by different informants and possibly both purposes are served by this act. The "true body" of the sun is then carried by N!adima to the far country in the east (their passage can be heard after dark; if one listens very carefully a swishing sound is audible in the sky) and it rises up into the sky the next morning, having regrown its "top body." The Milky Way is one of the paths along which N!adima carries the sun.

In winter N!adima moves the sun's daily course farther to the north, so that its heat will be lessened. The northerly declination means a lower altitude at zenith, and low altitude is equated with greater distance by analogy with the lower altitude of the rising and setting sun, when its heat is less. (The parallel case of a fire was offered in explanation; the farther one stands from a fire, the less is its heat.) In other words, the higher the sun, the closer it is to G/wi country and, therefore, the hotter the day. Conversely, the lower it is, the less is its heat.

The moon is a much less important entity. It is only accorded explicit recognition on the evening when the crescent moon first appears. The bones of a game animal are thrown toward it and the G/wi recite the formula "There are bones of meat, show us tomorrow to see well that we not wander and become lost. Let us be fat every day" (i.e., show us where there are plenty of food plants and game animals). Although this ceremony is believed to bring good luck in hunting and gathering, it is not seriously thought to ensure success and its recitation is often forgotten, without regret being shown over the omission.

The phases of the moon are named and coincide with new, crescent, full moon, and last quarter. The direction in which the horns of the crescent point is believed by some informants to be an omen of the fortunes of the month to come with south-pointing horns boding ill.

No great interest is shown in the stars. A few informants said that "old people" had told them that the stars were the fires of the dead and that meteorites were the grass torches that they carried from one shelter to the next as they visited among themselves, as the G/wi themselves do. This theory was not confirmed by any belief in the migration of the dead to the sky, and most informants stated that N!adima had created the stars when he made the universe and that nobody knew why he had done so. The evil that G//amama is believed to throw down onto G/wi women is embodied in little wooden slivers that, some say, resemble meteorites but are in some way distinctive and not to be confused with the latter.

Some constellations and stars are named:

n//abedzi (literally, giraffes, feminine – "they are big, like giraffes' eyes"); Southern Cross
khwe g≠ēi !ui (man shoots steenbok); Orion
!xais !hodi/wa (night's backbone); Milky Way
xwedzi (specific name); Pleiades
g≠eisa (female steenbok); Peacock
g≠eikxaoma (male steenbok); Altair
/edzini (firewood-finisher); Regulus (only sets when the firewood is finished)
/edzinig/wa (firewood-finisher-child); Arcturus
//xona (specific name); Canopus
g/aokhu (specific name); Sirius
!u:≠ono (morning star); Venus (unnamed when it appears as the evening star). No other planets are named

The G/wi are poor astronomers and do not have the elaborate imagery of the heavens of the Khoikhoi (see Shapera, 1930:413–418) and of some other Bushmen (see Bleek and Lloyd, 1911:73–98, and Marshall, 1975:153–159). They recognize the coincidence of the appearance of certain stars and constellations and the onset of the various seasons, but they use the stellar calendar only in the timing of the male initiation school; when Sirius appears in the evening sky in spring the men know it is time for the second phase of the school to begin. There is also the fallacious belief that all steenbok does give birth when the Pleiades are in the evening sky (i.e., early summer). Such cognitive dissonance is striking among hunters who dissect gravid does and encounter kids in all seasons of the year (see Smithers, 1971:221). The stars are not used as navigational aids; the G/wi travel at night only when the moonlight is bright enough for the normal daytime landmarks to be visible, and even then the journeys are short. As far as I could discover, stars are primarily used as atmospheric indicators that portend changes in the weather and as timekeepers at night. I was unable to determine which variables were taken into account. It seemed that rain was indicated by particularly crisp and bright stars – these conditions perhaps were brought about by inflows of cold, moist air at altitude and ahead of depressions. The night sky is certainly under the regular scrutiny that covers the rest of the environment, as was shown by the fact that satellites were invariably noticed in their passages. The appearance of Sputnik I in 1957 had occasioned some speculation. Some thought it to be an ill omen, but when nothing untoward happened in the community, sputnik and subsequent satellites were dismissed as irrelevant by that band. However, a series of mishaps suffered by the people at Tsxobe were associated these with satellites for some years. When more satellites were launched and several were to be seen each night, however, the association was abandoned and interest waned to a mere noting of passages.

The spectacular Ikeya-Seki 1965 VIII comet (a phenomenon for which the G/wi have no name) was feared when it first appeared in the early morning sky. The ≠xade people thought it would perhaps kill them. However, they reasoned, there was nothing they could do about it and went about their normal occasions. When the comet had disappeared and nothing had befallen them, they decided that it was not something dangerous. Nobody could remember having heard of a comet before, which means that the 1910 appearance of Halley's comet had been forgotten.

The reckoning of time

The passage of a period longer than about a month is measured in terms of the seasons of the year, as described earlier. A completed succession of these constitutes one *khudima* (perhaps better translated as "a twelve-month" rather than by the more precise term "year"). The number of these that have passed is reckoned in terms of elapsed winters. Periods of a few days are reckoned as the number of days that have passed; if the G/wi stock of numerals of one to three is exceeded, the number of days is counted off on the fingers. Although it would be theoretically possible to keep tally by means of this system of unnamed decimal-based numbers, people lose track of the days after a week or so and resort to qualitative reckoning (e.g., "a clump of days," "a big clump of days," "many days"). Although a lunar calendar is not explicitly referred to, pregnant women do keep track of the number of missed menstrual periods and correctly relate this to the season in which birth will occur. I found it convenient to refer to the phases of the moon to indicate to informants the timing of my own past and future movements, and they readily grasped and used this chronology. The principle therefore seems to be clear to them but is not lexically apparent.

The most frequently used time measurements are the diel divisions:

!u:!xaisa (morning-night); first light

ghiu//xa://xa: (burn [becomes] warm); dawn

//aba n/i ≠'kwa (light has emerged); sunrise

/ama n/i //o (sun has [become] warm); morning

/ama n/i //xa: (sun has [become] warm – warmer than //o); midmorning, forenoon

k'woni (specific term); midday

g!ua (specific term); midafternoon

g//ua (descend); late afternoon

es wa haiswa (she [the sun] is in the hole); sunset

es n/i ≠'he (she [the sun] has fallen); sunset

//haosa (specific term); dusk

!xais n/i ha: (night has come); late evening, until zodiacal light disappears

g//o: !xais (big, proper night); after about 10 P.M.

g/as:s kje (midnight stand, i.e., has arrived); midnight

!u:s n/i xao (morning has cut); false dawn; when the eastern horizon begins to lighten

!u:s n/i xao !xanakxi (morning has cut closed-face); when stars begin to fade

A G/wi camp never has an uninterrupted night's sleep. There is always someone awake, adding wood to the household fire, eating a snack, seeing to a child, listening to a strange noise in the bush, or keeping watch if dangerous animals are near. For this reason the divisions of the night are almost as important as those of the day when it comes to relating the events of the 24-hour period.

Time is conceptualized as a linear progression in which events are placed. Their occurrence is unique, that is, although there may be many ways in which one event resembles another, the same thing

111

never happens twice. To the extent that it differs from other events in its context, details, or other circumstances, each problem therefore requires a novel solution and each experience a fresh explanation.

The G/wi clearly take the future into account when planning their migrations and when noting the state of vegetation while moving around the country. These prognostications are based on annual cycles and do not extend beyond one such cycle. Longer perspectives of the future are within their mental powers, as they demonstrated when we discussed conservation, the probable consequences of replacing the bow and arrow with firearms and of allowing others to hunt in the Central Kalahari Game Reserve, and so on. The concept that a baby might grow and live to become a grandparent is also clear enough to them. However, the caprice of the next wet season precludes realistic looking ahead beyond one year at most, and in that sense the G/wi are oriented to the present and the near future in their practical planning. There is no discernible orientation toward anything resembling a conscious G/wi tradition, nor is there any chronicle of events beyond the recent past, which might provide dating data. Compared with Bantu in the Kalahari, the G/wi have little interest in the past; certainly there is nothing that could correspond with the Central Australian Aboriginal identification of the present with the past. With the above reservations, the G/wi are principally present-oriented.

The subterranean region

The region beneath the surface of the known world was also created by N!adima but is largely beyond man's ken. Its structure was described by an informant, "From this sand downward, it is soft, like the sand itself is. Then come stones – stones such as we do not see on the earth [in the known land]. Then, after those stones, comes water; water, water on, downward, endlessly. This is how it happened when the drilling-man drilled at /xo /'kei [i.e., the speaker saw the succession of structural materials reflected in the sandy, then rocky samples that came out of the borehole before water was struck]. But even before that, before the drilling, we knew how it was below because the 'old people' told us of it. They knew it must be so. They could see that there are some stones in the ground here [i.e., calcrete and diatomite]. They know there are other stones sitting in the ground [embedded in the surface] at G/axa and /ao [respectively, Dagga Camp and Sonop Hills]. They saw that water sinks into the ground when it rains and that it never comes up again."

The underworld, for which there is no specific term in G!wi but only a series of circumlocutions, is inhabited by *!!a:xudzi* (angry things, monsters). Nobody who has seen them has survived to describe them, so their appearance can only be guessed at. They only impinge on human lives when angered by the breaking of certain taboos. Cursing the land, setting fire to a shelter, excessive and malicious lying, murder, flagrant infidelity by young wives, and menstrual blood falling onto the ground will anger the *!!a:xudzi* and provoke them into coming to the surface and harming people. This they might do by spreading illness or by causing people to become blind or by killing people instantly and violently but without leaving any marks. Again they might do no more than give a man or woman a stiff and painful neck for a few days, or they may rob individuals of their senses, leaving them idiots. The monsters will wreak their damage and nothing can be done to stop them; they will only depart when satisfied. They might also express their anger by suppressing plant growth, according to some informants. The harm they do is not directed against the person breaking the taboo; it is wreaked upon those who encounter the monsters while they are on the surface. The minor ailments and illnesses that they bring can be treated with the normal medication appropriate to the symptoms. More serious attacks cannot be alleviated unless N!adima chooses to intervene.

The underworld is also the abode of *g!amadzi* (postmortem spirits), which are liberated from the body after a corpse has been in its grave long enough for the body to decay. (The status of the spirit during the individual's lifetime is unclear.) These spirits are intensely hostile to living persons, whom they will attack if given the chance. Their mobility is limited, but they will certainly attacked anybody who comes within a couple of hundred meters of a recent grave. The danger is greatest to those who loved and were loved by the deceased during his life. The victim does not die but is left blind and witless. Although informants did not state this, it seems that the postmortem spirit is lonely and captures the spirit of the living person as company, which suggests that the spirit is immanent during life. Because of this danger, graves are given a wide berth once the funeral ceremonies are completed (three days after interment), and the deceased's possessions are broken and left on top of the burial mound to warn travelers and others who may have strayed to keep away from the locality.

The desert G!wi do not share the belief of some ranch G!wi, the Nharo, and some other Bushmen that postmortem spirits eventually go to live in the sky, although there is an echo of this belief in the explanation, no longer accepted, that the stars are the fires of these "people." The desert G/wi did not, in my experience, ever use the word *khwena* (people) in reference to the postmortem spirits. Their belief is that *g/amadzi* remain forever in the underworld, gradually losing their power (as the deceased is forgotten?) until they become harmless. All spirits experience this fate, regardless of how they may have behaved while living above the earth.

The distinction between the *g/amadzi* concept and the Christian beliefs of eternal life with heavenly rewards or hellish punishment of the soul, according to the manner of life on earth, is significant evidence that G/wi theology is indigenous and not an attentuated form of the teachings of missionaries in Botswana and surrounding countries. The similarities of shared monotheism and the superficial parallel between Satan and G//amama are fortuitous coincidence.

Ontology

The G/wi view their environment, the knowable "middle world," as an unequivocally ordered world. Entities, events, and phenomena are actually, or potentially, explicable in terms of causal relationships with other entities and so on. These relationships are not chaotic or random but are regular and potentially repetitive. Caprice and irregularity are familiar complications but are seen as arising from changes in the factors comprising the relationships and not from suspension of, or departures from, order.

Gödel's incompleteness theorems state that in any complete formal axiomatic system there must be one or more propositions that can be neither proved nor disproved by the axioms and rules of inference of that system. Any logical statement or construct is necessarily founded upon a set of assumptions. The statement must be consistent with the assumptions, but given that consistency, the construct cannot be used to test the validity of the assumptions. Such testing can only proceed by means of recourse to some other appropriate construct or statement with its own underlying set of assumptions. Analysis of causal relationships, for instance, proceeds from one level to another in this fashion until either ultimate causation is discovered or the analysis is halted at the barrier of a set of assumptions that cannot be tested by the information or by any other means available to the analyzer. Rationality of thought is characterized by continued search for further levels of causation (i.e., for constructs to test the assumptions underlying the last-reached proposition). Conversely, rationality is abandoned when ultimate causation is immediately posited without attempting to find appropriate constructs, propositions, and so on. To illustrate this argument, a trite, but perhaps familiar, example may serve: If my children have the temerity to ask why they should keep quiet after I have roared at them to do so, I will have posited final causation by telling them, "Because Daddy damn well says so." It would be both more reasonable and rational of me to explain that I have an impatient editor waiting for a long-overdue manuscript and that their youthful cacophony interrupts the snail-like train of my thoughts. (In the unreal world of child-care books I could then go on to explain the necessity to publish lest all of us in the family perish and so on until long past their bedtime.) To ascribe the twitchings of the San Andreas fault to the Will of God is admirable piety, but in terms of this argument, it is essentially irrational. To invoke chance as the cause of the fault's convulsing just at the time that I invest my life's savings in real estate located along it is perhaps more fashionable but equally irrational.

In accounting for their experience of the events and phenomena of their world, the G/wi offer explanations that are, in these terms, essentially rational. Their analyses of causal relationships are halted by lack of knowledge or peter out at the boundary of relevance to their own lives but do not disappear down the hole of final causes.

The explanation of the chilly gust coming from the thunderstorm on a hot afternoon was a logical statement that rested on such assumptions as the ability of rain to displace the air through which it falls and on heat being transferred along the temperature gradient between ambient atmosphere and hail. The statement was consistent with the assumptions but could not be used to test their validity. That required recourse to, for instance, the concept of the storm-leopard, which is alien to my world view. But I cannot deny the rationality of this construct, which, in its different way, makes order of the apparently chaotic movements of the cloud mass. It would be silly of me to try to falsify the concept by pointing to a leopard on the ground and challenging an informant to raise it up thousands of meters into the cumulonimbus. He is not talking about that kind of leopard and would find my demand as comical as my proposal to "drink the rain." Instead, I must inquire further into the assumption, namely, the nature of the storm-leopard. Not unexpectedly, this line of investigation soon led to a loose strand. Nobody knew very much about storm-leopards and no one could tell me how they got into clouds; it was beyond the range of relevant knowledge. But it was all potentially explicable by the same mode of analysis. If, for instance, N!adima had put it there (which was probably the case), then the mechanics of this particular mode of cloud seeding would be no more mysterious than any of his other acts. His motives, although a mystery beyond human ken, would be in terms of familiar relationships of causation, for he is seen as a rational being who knows why he does things. (The G/wi seem to subscribe to what Salvador de Madriaga called "the relatively modest dogma that God is not mad.")

The structure of causal explanation is not linear, with each event being ascribed to a single cause, which, in turn, is the effect of a chain of one-to-one relationships. There is a web of causation. Factors combine in multiple causation and the strands merge where different events have in common one or more causal factors. The G/wi mode of explaining their world has much of the structuralist cast that Levi-Strauss (1966) found in the "untamed thought" of members of other small-scale societies. It seems to me that there are also interesting parallels with the constructs of von Bertalanffy (1968) and other general systems theorists. Structuralists and systems theorists have a common concern with the interrelationships of parts of the whole and with processes of self-regulation and transformation. The concepts of the systems theory of development and of entropy-negating processes in open systems have their counterpart in the G/wi account of the dynamics of their world.

It is axiomatic that a system consists of a set of interacting elements (von Bertalanffy, 1968:38) and that these elements should have in common characteristics that are relevant to the system as a whole (von Bertalanffy, 1968:54 ff.; Kremyanskiy, 1969:130 ff.). Although systems that are closed to, and impermeable by, their environment can be posited as a hypothetical type or heuristic device, real systems exhibit varying degrees of openness to, and communication with, their environments. Some measure of closedness is essential to the identity of a system; a wholly open system could not be detected (e.g., one could not photograph an invisible man) and would be unable to communicate by responding to, or impinging upon, its environment. Closedness is manifested in the system's capacity to interrupt, filter, or otherwise modify the inputs it receives from outside. These inputs constitute the energy and information by which the system maintains its operations and identity. Without them, the system will, sooner or later, succumb to entropic decay.

Among the array of objects and phenomena known to them, the G/wi conceive of some phenomena as being more immediately interrelated than are others (e.g., what we may term "the set": heart, blood vessels, and lungs). The conceived relationship involves inter-

116

dependence and interaction among the members of the set and corresponds to the concept of an open system. That the order, the organization, and the identity of a system is maintained by entropy-countering inputs of energy and information is mirrored by the G/wi view that there is an inherent tendency for things to fall apart, run down, or otherwise go awry unless there is appropriate interaction between systems that stand in the proper relationship. An obvious example is their contention that plants will wilt, die, or fail to reproduce unless sustained by rain in the right amount at the right time. Man and herbivorous animals are threatened by such failure of esculent plants and so on.

A cosmology that contains a concept of order must also include a concept and accommodation of disorder. Without at least the hypothetical possibility of disorder there would be no antithesis in terms of which order could be conceived and no contrast against which it could be discerned. Both Leach (1967) and Douglas (1970) have argued that departure from the order imposed or sanctioned by the deity may be interpreted as a change in the conceptual distance between man and deity. Certain disruptions of normal order render the individual closer to the deity in that the former acquires abnormal status or power (e.g., shamanistic trances; Leach's biblical example of Abraham's incestuous, but fruitful, intercourse with the aging Sarah). Other disruptions of order place the individual farther from the deity in that they induce a diminution of power or status, often as the consequence of condign sanctions that do not involve the direct intervention of the deity but are "built-in" as a sort of booby trap. Acts in the first category of departure from normality are only permitted (and are, therefore, successful) under prescribed circumstances. The point implicit in Douglas's (1970) thesis (and to some extent in Leach's) is that acts in this category – those that enhance power and/or status – are transcendental of the reigning order. Harmful acts, those regarded as evil or polluting, constitute a rejection or refutation of order. In short, order is identified with the deity and its denial is identified with evil.

Although N!adima himself is not regarded as the personification of goodness, that which is considered to be good is also supportive of order, and in this sense, order, goodness, and the deity are linked. Those of G//amama's actions that can be ambiguously interpreted (i.e., contrasted with the occasions when his agency is confused with, or perhaps seen as ancillary to, that of N!adima) are seen as disruptive denials of order. Unless he is frustrated or countered, the consequences of his actions are harmful and in this way there is a linking of G//amama, disorder, and evil. Any other agency that disrupts and denies order is similarly associated with evil. To put it in systems theory terms, evil and entropy are synonymous.

The "arrows" that are thrown down by G//amama to lodge in women induce entropy in the local social system. Trance dances are performed to correct this deviation from the steady state, and in transcending normal order in their trance states, the dancers restore the entropy in the system to zero or to a minus value. Negative entropy is drawn into the system by means of a transcendental act performed under prescribed circumstances. In similar fashion, in hunting and gathering, men and women bring food into their households and so maintain metabolic balance and resist entropic degradation. In so doing, they induce a measure of entropy in each of the systems from which they derive food. Although of a less dramatic nature than the previous example, this disruption of the normal order of food-producing systems can also be interpreted as transcendental action (see the following section, "Growth Processes and Degradation"), thus resolving the paradox implicit in the common status of all life-forms as N!adima's property, enjoying his protection: Failure to survive would entail diminution of N!adima's estate, but in order to survive, it is necessary to eat, which also diminishes the estate.

N!adima's actions, being those of the deity, are necessarily transcendental and, hence, negentropic and supportive of order. (Note that even N!adima's postcreation acts are restricted and do not achieve complete suspension of normal order.) For him to withhold rain will inhibit plant growth, disperse game herds, and possibly lead to the breakup of bands in the drought-affected area. However, he withholds rain as a sanction against human disruption of his order, and the long-term, wider-scale outcome of his action is to ensure conformity with his order by those who have been afflicted and to maintain order among those who have been impressed by his display of displeasure. With the idea of a negentropic N!adima, the uncertainty as to whether a life has been taken by him or by G//amama does not threaten the cognitive structure; rather than a whimsical deity, he is seen to be the one who will always maintain the order of his creation, even though his ways may be inscrutable.

Growth processes and degradation

Perturbation of the system's equilibrium that cannot be countered by the processes of steady-state regulation must lead either to degradation or to growth as the entropy induced by the perturbation either increases or is resisted. Degradation, unless countered by appropriate inputs, results in the death of the system. Growth processes require increased inputs to meet the increased demand of negentropy.

The arrest of entropic degradation by appropriate inputs is clearly conceptualized in the example of the Glwi version of the working of the respiratory-cardiovascular system and in the explanation of the food requirements of living things (see earlier in this chapter, under "Mammals"). This is the normal, maintenance type of requirement described above. Where abnormal perturbation has disrupted or exceeded the system's capacity for self-regulation and maintenance, abnormal inputs are required. In the treatment of some ailments, and sometimes to assist dancers in recovering from their trances, the Glwi practitioner rubs his own underarm sweat onto his palms and presses them against the head of the patient or dancer. This process is seen as something akin to a transfusion and is believed to (and evidently does) restore the person receiving the treatment. Medicinal herbs are seen as possessing a comparable power; the herb has this power as one of its properties, which are entirely mechanical in action, as are its fragrance or color. There is no distinction drawn between the nature of such herbal remedies for illness and those remedies that correct other aberrations such as persistent failure in hunting. Their power to correct specific departures from normality or a range of such disruptions is something that was included in their character when N!adima created the species concerned – but man had to discover that this power existed and how to tap it as the means of supplementing his own ability to resist entropy.

As Douglas (1970) has pointed out, menstruation (particularly menarche), birth, and death are commonly regarded as polluting events. To the Glwi, the sight of a girl secluded at menarche, menstrual blood, or a placenta will blind a man; contact with these will kill him. The husband of a girl who menstruates for the first time is endangered by her state and will do no hunting until her menarchial ceremonies are completed. No menstruating woman can touch hunting weapons without comparably endangering their owner, rendering him highly prone to serious accident if he subsequently uses them. Menstruation, birth, and death (see the earlier section "The Subterranean Region" on the dangers that postmortem spirits present) are associated with changes of state and are, in a sense, charged with entropy that must either be contained by the proper disposal of placenta, menstrual blood, and corpses (in which the disruptive power is held) or in the case of menarche by an equilibrium-restoring ritual involving "transfusion" of negentropy from other members of the band and the use of medicinal herbs.

In the G/wi view all life-forms are subject to environmental stresses that can only be countered by drawing upon environmental resources. Furthermore, all forms of reproduction require an increase in the resources above the normal requirements of the reproducing system. Plant reproduction is clearly conceived of in terms of seed production and seed vivification, which cannot take place without rain (which represents an addition to the plant's life-maintaining resources during the other seasons of the year). A band that splits into two after outgrowing its territorial resources clearly requires additional territory for the daughter band. But growth processes are limited in their extent by the resources on which the system is able to draw. A married couple should not, in the vernacular ideal, have another child before a previous one is weaned and should not have more children than they are capable of providing for (common opinion favors a maximum of four children, five being considered too many). The concepts of a limit to growth and of optimal size are fairly clearly formulated in the G/wi world view. During the spectacular displays of summer storms, lightning strikes are common; informants point out that it is the taller trees that are struck and killed – "they have grown big enough." Sporadic plagues of armored ground crickets attract flocks of ibises and other insectivorous birds. Informants commented that it was good that the birds came in such large numbers and ate so many crickets "or the whole world will be [nothing but] crickets." Inquiring further, I was told that this applies to all forms of life, "Everything must grow. It must bear its young. It must live. But not so that anything becomes too numerous in one place and takes everything" (i.e., leaves no resources for other animals or plants). Neither, however, should any species be depleted below its viable limit. This value is also expressed in not stripping an area of its plant resources but always leaving some specimens to reproduce and regenerate the local population. To do otherwise would be an affront to N!adima.

The world, then, is seen as a gigantic version of a self-regulating life-support system in which component subsystems, themselves self-regulating within certain limits of tolerance, interact to correct perturbations. This system is an open one; N!adima feeds in his "inputs," manipulating the subsystems and their elements to maintain the conceptualized equilibrium. There is constant interaction and motion, local and temporary changes occur, but the overall character and identity of the system continues unchanged – N!adima's universe as he created it. On this larger scale the G/wi concept of order is more clearly like the steady state of systems theory, in which the ratios and relative positions of components tend to constancy (Katz and Kahn, 1969:96–99).

N!adima created the context in which man lives but did not ordain precisely the manner of his living. Within the constraints of the other systems and their traffic, man must devise his own system, drawing from the flow but never breaking it. The G/wi see their culture as plastic; old knowledge is lost and new discoveries take its place. There is more than one way of doing something, more than one way of solving a problem, and more than one problem in a situation. There are many ways in which man may wire the circuits of his own system, and there exist alternative links by which it may be integrated into the network of the middle world. New means can be discovered to attain the same ends, or the same means may be used to attain new ends. There is in G/wi cosmology recognition of the principle of equifinality (see von Bertalanffy, 1968:131–134).

The knowledge that the G/wi possess at any stage must be at least sufficient for them to understand the structures and functions of the systems with which they interact and for them to be able to fit into the patterns of interrelationship among those systems. The only knowledge available to an individual is the knowledge that is stored in his and his acquaintances' memories. The total amount of information required for bare survival under central Kalahari conditions is considerable; that needed for competent participation in the sociocultural system of the G/wi is vast – more than the average anthropologist could acquire in many years of fieldwork. Some economy must necessarily be achieved to lessen the memory load. One of the functions of the universality of the basic pattern of systems in the cognitive structures of the G/wi is to achieve this type of economy. The effect of this common pattern, this isomorphism among conceptualized systems, is to focus attention on the similarities in the structures of the systems that are perceived as being isomorphic and to highlight those types of relationships that are common to the systems concerned, leaving only detail as an additional burden to the memory (see Levi-Strauss, 1966:ch. 1; von Bertalanffy, 1968:chs. 8 and 9).

The fact that the G/wi use the same anatomical terms for man as for other mammals does not mean that in G/wi eyes all mammals have the same anatomy. Rather, it is that the G/wi have discerned similarities in such aspects as the linear sequence of arrangement, spatial distribution within the body, and appearance and function (more obvious in the case of external anatomy). To give to the livers of all mammals the name *gēi'sa* is more economical than to give a different name to that organ in each mammal species. Where it is necessary to distinguish among them (e.g., to contrast the flat taste of wildebeest liver with that of the delicately piquant steenbok liver), identifying the respective species differentiates between the two. Levi-Strauss (1966:ch. 1) see systems isomorphism in cognitive structures as a universal human characteristic. I have no quarrel with this statement but emphasize that the extent of isomorphism in any one cognitive structure must depend on a minimal amount of discovery of characteristics common to the systems conceived of as being isomorphic. This is not, therefore, a primitive characteristic in the sense that it dates back to the beginnings of human knowledge (although it may well mark an early stage in the development of human understanding).

If the extent to which similarities have been discerned among systems limits the extent of systems isomorphism – as von Bertalanffy (1968:chs. 8 and 9) suggests – it does not follow that all systems to which the G/wi have given approximately the same close attention will necessarily display the same type of isomorphism. It would, for instance, be tritely ridiculous to expect an anthropomorphic floristic ethology comparable with the mammalian and avian ethology that the G/wi have developed. Isomorphism between the conceptualized mammalian and floristic systems is about another axis and is seen in the use of kinship models to express perceived resemblances between plant species and in the similar basis of nomenclature (specific names for the more significant species as against derived names for those species that are less important): the similarities in the notions of man's correct behavior in not disrupting the equilibria of the systems while exploiting them – that is, in his relationships with them. The same is true of the avian and invertebrate systems; the difference lies in the relatively small area of uncertainty in the conceptualized mammalian and floristic systems contrasted with the greater proportion of uncertainty in the avian system and much greater uncertainty in the invertebrate system. In these cases there are more generic terms, more unnamed species, and among invertebrates, the number of unknown species exceeds the known. Birds and invertebrates are more peripheral to man's interests than are plants and mammals, and there is less interaction with man that is perceived as relevant. The conceptualized structures are therefore less completely and closely integrated than are those pertaining to plants and mammals. So, although isomorphism exists, it is that between loose and sketchy structures.

122

It must be noted that the tendency in G/wi zoological and botanical nomenclature toward the specific naming of important species and the giving of derived names to the less significant species is a common phenomenon. Comparing G/wi and Tswana (i.e., those Tswana dialects spoken in and on the borders of the Kalahari), the two nomenclatures show virtual coincidence of the distribution of specific, derived, and generic names for mammals (which have the same relative importance for both G/wi and Tswana). The trend is as marked in the names of other animals and of plants, but the coincidence of distribution is lessened by shifts in relative significance. A comparable trend can be discerned in the English and Afrikaans names of southern African plants and animals and also in the English names of Australian plants and animals, although this is obscured to some extent by the historical factor and the global spread of original denominative species.

The covert taxonomy of the G/wi is based on multiplex cross-reference. For instance, members of the family of Curcurbitaceae are likened to one another in such broad detail as their common prostrate habit and in finer detail as their spirally coiled tendrils and the similarities in their flowers and fruits. *Harpagophytum procumbens* and *Dicerocaryum zanguebarium*, both creepers, are likened to the curcurbits in their habit but are associated with several other members of their family (Pedaliaceae) on the basis of flower similarities and to *Tribulus* and even to *Acacias* and so on in their thorniness. It is apparent that the G/wi have a fluid, complex set of characteristics with which they can systematize in setting up cross-referenced taxonomies.

Language

Language is not, of course, the only form of communication in which thoughts are expressed. There are also gestures, facial expressions, hand signals, and so on. But language is the most versatile and flexible method of communication and carries the greatest variety of content of meaning. In a speech community whose members share a common background of culture and experience, the language will inevitably reflect the patterns of thought characteristic of that community. What is debatable is the way in which these patterns are reflected. Edward Sapir (1966:128) stated:

Language is not merely a more or less systematic inventory of the various items of experience which seem relevant to the individual, as is so often naïvely assumed, but is also a self-contained creative symbolic organization, which not only refers to experience largely acquired without its help, but actually defines experience for us by reason of its formal completeness and because of our unconscious projection of its implicit expectations into the field of experience.

Whorf, who had studied under Sapir, explored this idea (see Carroll, 1956) and concluded that there is a relationship between habitual thought and behavior and language, a concept that became known as the Sapir-Whorf hypothesis. Whorf emphasized that "there are connections but not correlations or diagnostic correspondences between cultural norms and linguistic patterns" (Carroll, 1956:159) but did not clarify what the nature of the connections actually is. The idea is tantalizing; as Carroll (1956:29) says in his introduction to a collection of Whorf's papers, "Whether or not, in fact, we assume any mental processes standing behind them, we are led to put a high value on verbal responses in their manifold forms as the chief data relating to perception and cognition." But as Malinowski (1923, 1935) pointed out, language is but one mode of socially cooperative activity and cannot stand abstracted from its context as a "countersign of thought." Language has meaning within situations and in personal, cultural, historical, and physical settings.

It seems, then, that the cognitive structure of a sociocultural system can be predicted from the language only within narrow limits and, at least at present, the results will be trite. Levi-Strauss (1966:chs. 2 and 5) has demonstrated that cognitive structures are systematic and that any one structure consists of a series of transformations of a set of principles or premises. He has also demonstrated (although not stated in these terms) that the principle of equifinality operates in cognitive systems – that the same means may be employed to achieve different ends and that the same end can be achieved by different means. The existence of homophones and synonyms illustrates the operation of this same principle in language systems. Thus, even where the relationship between a structural feature of a language and part of the cognitive structure of the corresponding culture has been discerned, there is no certainty that a different but analogous linguistic feature will indicate a similarly analogous portion of the cognitive structure.

To resort to metaphor, if the cognitive structure and the language stood in an arithmetic relationship as in a series, thus

$$
\begin{array}{cc}
 & 2:4 \\
\text{as} & 4:8 \\
\text{and as} & 200:400, \text{ etc.}
\end{array}
$$

If the left-hand column represents linguistic data, the cognitive code in the right-hand column can be cracked once any one of the relationships has been established.

The transformations are, however, more variable than this and could be represented in the metaphor:

$$1:A$$
$$\text{as} \quad 2:B$$
$$\text{and as} \quad 3:C \ldots 26:Z.$$

Now, in such books as Grace Gabler's *Child's Alphabet* (1945), the letter *A* is accompanied by the printed words for, and drawings of, ape, alligator, ark, and acorn. The value 1, therefore could be represented by a drawing of any one of these four objects. Similarly, the value 2 could be represented by a drawing of a bird, ballerina, butterfly, or boat. The relationships are systematic in the way described by Sapir and Whorf, but the relata are unpredictable except in, first, a negative sense (e.g., 3 will not be represented by a daisy, drum, or duck, etc.) and, second, in the very general terms that 19 will be represented by any one of the thousands of English words beginning with the letter *S*. Even then, there is uncertainty – a sketch of an alligator may well represent 3 to African children who are more familiar with crocodiles.

An example of the variability of linguistic expression of cognition is the English vocabulary used to describe the taste of wines. This includes tactile metaphors (rough, smooth, warm, cool, etc.), some of which reflect particular attributes of substances far removed from the experience that they are being employed to communicate (e.g., silk, velvet). Other metaphors may be derived from human personality attributes (pretentious, friendly, etc.). The analogies underlying the metaphors are consistent in that

attributes of *wine A* : attributes of *wine B*
as sensory experience of *warm* : experience of *cool* . . .

and so on, through the connoisseur's terminology, but the relationship of silk:velvet is not isomorphous with the relationships of rough:smooth, nor of pretentious:friendly. Such isomorphism of the sets is necessary to the validity of the Sapir-Whorf hypothesis and to the possibility of prediction of the nature of one on the basis of the nature of another.

In short, there is "fit" between the linguistic and cognitive structures, but the extent and topology of that fit can only be predicted in rather general terms. The significant relationships of which Whorf speaks can only mold thought where is some causal link that can be demonstrated; necessity and sufficiency obtain only in relatively rare instances. This chapter concerns the G/wi view of their environment. What follows is a discussion of some of the features of the language and the extent to, and manner in, which these characteristics express certain views and interpretations of the environment. A small number of examples of Whorf's significant relationships are given, but for the reasons explained above, a linguistic description cannot provide the key to the cognitive structures of the G/wi.

A noun in G/wi consists of a noun stem and a suffix that indicates number and gender. There are three grammatical numbers, singular (one object, phenomenon, or whatever is being named), dual (two objects), and plural (three or more objects). There are also three genders, masculine, feminine, and agglomerative. Only nouns of dual and plural number have agglomerative gender. There is no article in G/wi.

The content of the masculine gender is:

1. Humans and animals of discernible male sex
2. Long, tall, elongated, sharp, or narrow objects
3. The right half of the body, its limbs and objects associated with them (e.g., the right sandal)
4. Meteorological phenomena, including wind, rain, lightning, and thunder in, or coming from, the northeast and northwest quadrants, and driving rain; the waxing moon or other harmless, fruitful, or beneficient natural phenomena
5. Any single person or animal, the sex of which is unknown
6. Names of men and boys and of N!adima and G//amama
7. The caducous parts of flowers
8. Rifle and shotgun cartridges (if No. 2 above is interpreted as having phallic significance or association, the ascription of male gender to ammunition is consistent. The cartridge penetrates the breech of the rifle or gun)

The content of the feminine gender is:

1. Humans and animals of discernible female sex
2. Short, round, blunt, wide objects.
3. The left side of the body, its limbs and associated objects (e.g., the left sandal)
4. Meteorological phenomena in, or coming from, the south, intermittent rain and drizzle; the sun and the waning moon and other hot or cold, barren or destructive natural phenomena
5. Any large number of animals or objects in one group, plural number being used
6. Fluids and anything composed of very many small particles (e.g., sand, powder, tobacco); widespread drizzling rain

126

7. Seasons of the year
8. Names of women and girls; the wife of N!adima
9. The seed-forming parts of flowers
10. Firearms
11. Metals in any state previous to being made into G/wi artifacts (including lengths of wire, which, in terms of the masculine No. 2 might be classified as masculine)

The content of the agglomerative gender is:

1. Aggregations (dual or plural in number) of persons, animals, or objects of masculine and feminine gender
2. The generic or group names of animals or people.

Noun suffixes of the singular number and the plural agglomerative gender distinguish the subject/object status of the noun in the utterance. A full range of pronouns distinguishes person ("I," "you," "they," etc.), number, gender, subjectival-objectival status, and inclusion-exclusion in the action of the principal clause of the utterance.

Nominals (i.e., nouns and pronouns) are qualified by a series of adjectival modifers that are structurally similar or nearly so. These carry the semantic significance of adjectives, interrogatives, relative constructions (e.g., "that which"), and demonstratives ("this," "that," etc.). These last are very elaborate and four positions are distinguished:

1. In the speaker's area or next to him
2. In the listener's area, between him and the speaker
3. Beyond but near the listener
4. Beyond and far from the listener

The series also indicates whether or not the position referred to is known to the listener.

There are several nonadjectival noun constructions that have adverbial or quasi-adverbial function, including locative ("at," "on," "to," "from," etc.), instrumentative ("by means of"), associative ("together with"), and comparative forms.

Verbs are modified adverbially and by a variety of suffixial extensions, which give the meaning of the passive voice, neuter ("to become" – the subject of the verb enters into some state or condition without stipulating the agent of the action), causative ("to cause to do"), reciprocal ("to do to one another"), intensive-extensive and repetitive action. There is an extensive range of temporal adverbs that function as tenses and also other adverbs of time, manner, duration, and intensity. One of the verbal extensions also functions as a tense, the perfect stative. Many noun and verb stems are given adverbial function by the suffix -si.

A copulative series ("it is," "they are") is the only grammatical feature in which women's speech differs from that of men. The series indicates subjectival or objectival status and includes interrogative ("is it?") and adverbial ("it is hot") forms.

There are negative forms of all verbal and quasi-verbal constructions.

A speaker within his own social circle is permitted great latitude in style and syntax. Such elements as noun-gender-number suffixes and tense formatives are freely omitted. Once a factor (e.g., gender, number, tense, the subject of the actions, the place of action) is established, it may be left out of subsequent utterances to which it is relevant. Establishment of such factors may be by nonlinguistic communication (e.g., gesture), or the factor may be known to both the speaker and his audience. Conversation can thus be rendered almost incomprehensible to an outsider and totally beyond the comprehension of one whose command of the language is poor. I have occasionally seen baffled native speakers request elucidation.

Word order, with a few exceptions, is highly flexible and in the absence of information contained in the omitted elements adds greatly to the difficulty of following a conversation. Verbs may precede or follow nominals, and descriptives seldom have a fixed place, for example:

> *kjidi n/i twē:s g//eikwesa m:*
> (I did-today pretty female-person see)
>
> *g//eikhwesa twē:s kjidi n/i m:*
> (Female-person pretty I did-today see)
>
> *g//ei twē: kjidi m:*
> (Female pretty I see)

These variations were given to me as equivalent utterances. The G/wi are sometimes given to considerable repetition, in which they will reiterate the same set of facts by this type of paraphrasing, thus ringing the syntactical changes and driving a point well home.

In contrast to this reiterative style, a highly allusive somewhat covert style of narration is also employed. The example shown in Table 6 is taken from an informant who was relating a series of incidents, some of which I had witnessed. It is by no means an extreme example of the omission of language elements and of what most English speakers would probably regard as information essential to compre-

Table 6. *Example of Gǀwi narrative style*

nǁam — Then
!uã — carry-home
(Then we men and women carried home the meat of the eland bull

nǁakim — this-one-referred-to
itse — we-two-men
kxo: — eat)
and we men ate of the meat of the eland bull referred to.)

abe — (He
khwe — long-ago
ama — him
ǁao — shoot-at)
(He shot at the eland bull a long time before this.)

abe — (He
ma — him
ǁao — shoot-at
ama — him
!ui — hit)
(He shot at the eland bull and hit it.)

abe — (He
ǀxei — fall
kji: — not [excl.]
igǁei — we-men [plural excl.]
!hao — chased-after)
(The eland did not fall down, so only we men [i.e., not including you, the listener] chased after it.)

atsera — (They-two-men [inclusive]
kxẽida — vulture
xu: — come-from
ha — come)
(Then you [the listener] and the other man came back in the aircraft and came to us.)

agǁei — (We-men [plural inclusive]
khuma — next-day
nǁam — then
am — his
nǁaka — horn
u: — bring)
(You came with us men the next day and we brought the eland horns home with us.)

hension. The mixed order of events is by no means unusual in G/wi narrative.

Where speech elements are included in utterances, the subject/object relationship of singular and agglomerative plural nouns is clearly established by their number-gender suffixes. The most common pattern of the active predicative is

subject	object	verb
ababe	*/xo:ma*	*kji m:*
(Dog	gemsbok	is seeing)

This may be reversed to

object	subject	verb
/xaoma	*dzerasi*	*n/i be:*
(The bird fled from the snake today)

Syntactical fluidity can lead to ambiguity where noun-gender suffixes do not distinguish subjectival/objectival status; for example:

> *n!unidzi g!we:g//wa tju pa:*
> (Rats/mice bullfrogs [plural masculine] yesterday bite)

Here it is not clear whether the rats/mice were the biters or the bitten —the probabilities are equal.

The number system of the nominal structure imposes on the speaker the obligatory observation of that which is named by the noun or pronoun as to whether it refers to one, two, or more objects. This suffixal indication of grammatical numbers, of which two coincide with arithmetic numbers, makes the first two cardinal numbers in the vocabulary virtually redundant for counting purposes. Only the remaining cardinal number, *n!ona-* (three), is essential in counting. In practice, the first two cardinals, */wi-* (one) and */am-* (two), are used only for emphasis of arithmetic number. The paucity of numerals is no handicap to the G/wi, who seldom need to count; when they do, they manage adequately by indicating unnamed arithmetic numbers on their fingers. The totals or amounts of objects are more often of importance (e.g., the food-plant harvest of the day; the size of a herd of game animals; the population of a camp), but these totals are not assessed as numbers but as amounts, that is, qualitatively rather than quantitatively.

The gender system requires the speaker to classify that which he names according to the criteria of the gender categories (i.e., shape, biological sex, etc.). The feminine gender includes females and round, short, or squat objects. This contrasts with the masculine

130

gender, which includes males and elongated objects (which might be regarded as a moment of phallic symbolism). In regard to natural phenomena, the beneficent, life-giving, or superior rains or winds are of masculine gender. Phenomena that are harmful, barren, or inferior (e.g., the sun, the cold, and the light, early rains) are feminine in gender. Among these almost entirely right-handed people, the right side is of masculine gender and the less-used left side has feminine gender. This might seem tempting evidence for a hypothesis that a concept of male superiority is being expressed in the gender system. Informants, however, rejected this suggestion and denied that the inferior criteria of dangerous, barren, discomforting, and so on have anything to do with the inherent characteristics of women or that the superior criteria reflect male superiority. The gender system is, in fact, a means of classifying contrasting criteria, the various sets of which do not necessarily bear any relationship to one another. If the possession of characteristics A, B, and C identifies objects as having masculine gender and D, E, and F identify objects as feminine in gender, there is not necessarily any factor common to A, B, and C (or to D, E, and F), but A contrasts with D, B with E, and C with F.

Criteria that are meaningful to hunter-gatherers determine the content of the two principal genders. The biological sex of humans is socially and economically significant and the sex of many animals determines such features as size, food value, and behavior. The shape of specimens of vegetation often determines or reflects their economic value; the shape of fruits, roots, and bulbs is closely related to their food and fluid content. The meteorological criteria are self-evident.

The ascription of masculine identity to the deity cannot be explained in terms of G/wi men having a status superior to that of women.

The agglomerative gender is not the label of intermediate, or noncontrasting, qualities but is used to indicate pairs or larger groups of objects, each of which has separate feminine or masculine gender. For this reason, the term "agglomerative" has been used in preference to a term such as *communis*, or common gender, which implies a position intermediate between the masculine and the feminine gender.

The complexity of the pronominal structure, which is employed mainly in the human context, requires considerable precision of reference. It requires a knowledge of the sex and the number of persons referred to and may also necessitate specifying their relationship to the action mentioned in the verb. This imposes a keen awareness of social relationships and the activities of other people.

The relative complexity of the noun structure allows some versatility in the use of noun stems, increasing their semantic utility and permitting an economy in their total number. This is borne out by the fact that the natural environment has a number of objects that share noun stems. I suggest that this indicates an awareness of, or at least a concept of, the relationships between objects that are named by the same noun stem and their ability to have some contrasting qualities when accorded different genders. Where this synonymity is applied to raw materials and derived artifact, it argues against the view that the G/wi feel that the process of manufacture alters the essential identity and nature of the material used. Thus the act of manufacture would not sunder the link between an artifact and its origins. Such a view is consistent with the simple technology of the people, in which no manufacturing process is sufficiently complex or drawn out for such a link to disappear (i.e., as distinct from the understandable view that an automobile is very remote from the tons of iron ore that yielded the steel from which much of it is made). It could be argued that this linguistic persistence of identity reflects the awareness of the environmental system from which the materials are drawn.

The subject/object system, although rudimentary and not developed in the whole of the nominal structure, indicates a clear understanding of agency, and where it operates, it imposes on the speaker a mandatory observation of agency.

The absence of any structural distinction of human versus nonhuman and animate versus inanimate and the existence of two genders of equal status (with a third that is a symmetrical combination of them) indicate that all objects are considered to have a measure of similarity. Even if it might make cultural nonsense, it would be structurally possible to express the idea that a stone might possess some of the characteristics of a man, that is, the language permits exploration of this concept. Consequently, the language permits free comparisons, the formation of analogies, and the expression of the isomorphism of systems.

The structural relationships between adjectives and other nominal descriptives make it clear that the qualities or modifications expressed by the descriptives are characteristics of the things that are named by the nominals.

The parallel series of demonstratives require that the speaker be aware of whether or not his listener knows the position of that to which he refers. Refinements of the third-position demonstratives are orientated in terms of the listener. Such consideration for the listener requires a measure of social sensitivity and awareness and makes it explicit in the utterance.

The absence of a structural device for expressing elaborately ranked comparisons (such as the degrees of comparison in Indo-European languages) is perhaps a function of the relatively small amount of choice allowed by the environment and the consequent futility and irrelevance of such hierarchies and, furthermore, is in accord with the egalitarian social organization in which comparisons seldom occur. There is, however, a simple, direct comparison device that expresses the sort of choice exercised in selecting from a range of possible, but ranked, hunting targets, migration routes, destinations, and so on.

The temporal modifications of the verb indicate a clear relationship of event to time. The divisions of time of action are essentially future, present, and past. There is a notable clustering of divisions about the present (i.e., the future, three present, hodiurnal and hesternal divisions of time). This accords with the relative indifference of the G/wi to time remote from the present. In a subsistence economy, where it is almost impossible to store food for longer than a day or two, it is understandable that there should be no exact structural distinction of events more remote than two days. It may, of course, be argued that English does not define time any more exactly in its tenses, but this lack is absorbed by a wealth of adverbial and quasi-adverbial time indicators of fine precision (e.g., reckoning in milliseconds).

The verbal extensions indicate an understanding and concept of the varieties of process and action and of the close identification of action with the circumstances that surround it. These extensions also reveal a certain flexibility of conceptualization of agency.

The wealth of compound verbs (i.e., verbs compounded of two or more verb stems of different significance to express a new or different action or process) shows creative imagination at work in conceptualizing the relationship between the actions or processes whose verbs are thus compounded and an appreciation of the meaning of the zone of overlap of their two apparently discrete areas of significance. It is emphasized that a compound becomes a single verb and is not a sequence of verbs. Semantically, too, the compound may express processes or actions that are not obviously related to those of the constituent verb stems: for example, ≠e:n!wa:xo (to apportion meat), from ≠e: (to flay or skin a carcass), n!wa:xo (to place, put); //uma/xei (to fall asleep), from //uma (to sleep), /xei (to fall). The latter example is a coincidental convergence of G/wi and English conceptualization.

Although their language indicates that they view object and action as different orders, it is clear that they do not consider the latter except as existing through the agency of objects, that is, objects exist independently but processes and actions can only take place through the agency of objects. Even those utterances that consist of verb stems only (e.g., imperatives) imply the existence of a subject to whom the imperative is addressed. It is true that there is considerable interchange of the qualities of objects and actions in the versatility of descriptives and that there is structural transfer of stems between nominals and verbals, but this does not imply that the two orders, object and action, are merged or confused.

The G/wi vocabulary is rich in environmental nomenclature but not exhaustive. Features of the natural, social, or supernatural environment that are either glaringly self-evident (e.g., the family) or not seen as meaningful or important are often left unnamed. The dearth of vertical or hierarchical taxonomic labels can perhaps be explained by the relatively small number of named life-forms. Taxonomy is governed by the existence of common characteristics, and the process consists of grouping together in one class those individuals that are perceived to have qualities in common. Hierarchical taxonomies are capable of being almost indefinitely expanded in the form of a geometrical progression, but horizontal taxonomies have a much smaller capacity for expansion. I suggest that the need for vertical or hierarchical taxonomic labeling does not arise where the stock of named species is small enough and the overt taxonomy can, instead, develop in the horizontal dimension. In G/wi this development is represented by such devices as homologue and unspeciated generic terms. For the rest, the number of individually named species is small enough to permit their comfortable accommodation in the memory. A comparable situation is found in the labeling of entities in the social environment. The personal names of individuals usually commemorate events or circumstances connected with or preceding their birth and synonymy is therefore rare. The social environment of a person is limited to the 250-odd other members of his own and allied bands. Their names are remembered with no apparent difficulty and are a sufficient means of identification. There are no names of social groupings corresponding to "household," "family," "band," or "tribe" that might be added to the personal name as a taxonomic device (except insofar as the associative plural may function in this manner). Although kinship terminology is not extensive, it is adequate for G/wi needs in that all recognized kin can be identified without intolerable anomalies or confusion. There are some social concepts included in the vocabulary for which the G/wi have no cultural equivalent, for example, //xeixama (chief).

Collective nouns indicating aggregations of objects (e.g., "flock," "herd," "pride") do not occur. This reflects the hunter-gatherer's preoccupation with quantity. Although this statement appears to contradict what I have said about numerals, collective nouns are too imprecise for the G/wi, who require information on whether the animals are single, two, or more. If more, it would be of little importance if the number were greater than ten, which can be readily indicated on the fingers of two hands. The point is that there is a very limited range within which exact numerical information has relevance to hunting and gathering, and the G/wi verbal and nonverbal system of counting is adequate within this range; consequently, the information contained in a collective noun would be redundant or valueless.

The lack of distinction between the young of man, animals, and plants probably reflects the hunter-gatherer's concern with size as a determinant of economic value. Unlike the herdsman, who is concerned with the physical maturity and productive capacity of his animals, the hunter can equate youth and smallness as far as food yields are concerned.

Onomatopoeic nouns are virtually confined to the names of birds whose calls they echo, indicating that sound as a distinguishing characteristic is not accorded the same importance as it is by some other people (e.g., the English who have a much richer stock of onomatopoeia). The central Kalahari, of course, has fewer sounds than are heard in an English wood or city, but it still seems to me remarkable that there are not more onomatopoeic names, particularly of insects, as the calls of many of these are their most obvious distinguishing characteristic (e.g., crickets and cicadas).

Compound nouns are common. In many instances the formation of the compound involves the conceptualization of a functional connection between two discrete objects in the labeling of a third (e.g., *dju:khosa*, literally, rain-skin, cloud of cumulus and/or nimbus families). Other compounds are formed by combining verb and noun stems, for example, *pa:xusa* (literally, bite-thing, carnivore, or any biting, stinging fauna). Many personal names are compounds of this type. This formation indicates an association of action and object in which the object is distinguished by its action.

Verbs occur more frequently in texts, which I have collected, than do any other speech elements (verb frequency averages 50 percent of content). In spite of the fact that the verbal genera have a greater number of constituent species (e.g., the variety of extensions) than do the nominal genera (which are limited to number, gender, and subject/object status), there is also a greater proliferation of verb stems than of noun stems. This is perhaps a reflection of the comparatively uniform environment, the paucity of objects of material culture to which specific names (i.e., as distinct from names synonymous with, or derived from, the parent material of the artifacts) have been given, and the small variety of artifacts that comprise the material culture. The significant variation of verb stems is quite as great in G/wi as it is in English, Tswana, or Afrikaans. In some fields of activity, for example, cooking and hunting, there are many different actions that are distinguished with narrow precision. Other aspects of G/wi life are less well served by verbs, which have wide areas of significance. I was not able to discern any pattern in this variation. For instance, the vocabularies of affection and disaffection are equally brief and wide in scope: *wi:* (to like, be fond of, love, etc.) and *//xa:* (to be displeased with, annoyed by, angry with, furious, to be fierce, to argue with, to fight with, etc.).

There are few onomatopoeic verb stems and most are echoic of human sounds, for example, *kx'aije* (to belch), *nh/hudu* (to hiccup), *!xhunu* (to snore), again indicating the relative unimportance of sound to the G/wi in their quiet environment.

The structural flexibility of association of object with action allowed me to use the observed range of verbs that are meaningfully associated with particular nouns as an indication of the qualities and abilities that the G/wi ascribe to those objects and, hence, are an indication of certain aspects of their world view.

The great flexibility and variation of word order and the extent to which speech elements can be omitted from utterances reduce the amount of information that is clearly and unambiguously contained in some utterances. This can only be tolerated in a situation where information is contained in the nonformal aspects of utterances, in nonverbal form, or is part of knowledge common to speaker and listener. Although the G/wi make extensive use of nonformal aspects, such as cadence, pitch, and emphasis, and of nonverbal devices, such as gesture, to supplement the formal content of their utterances, the bulk of information omitted from the formal content of the utterance is that which is already known to both speaker and audience. The brief example given of the amount of information that may be left out gives some impression of the extent of that shared knowledge. This high incidence of shared items of information can only occur in a closely knit community whose members have a sufficient degree of mutual concern to note and remember one another's sayings and doings.

At the other end of the scale is the highly repetitive style of speech in which utterances are paraphrased again and again without significant addition to information content. This style of speech, which alternates freely with the cryptic, allusive, elliptical style, is usually accompanied by the audience chorusing the last syllable of the speaker's "sentence." There is clearly no semantic information contained in the repetition and chorus, but from the facial and other expressions of the audience it is evident that this repetition achieves and expresses an emotional communion between all those partaking in the conversation, a feature that is also indicative of the intensity of G/wi interest in the social environment.

4

Social Organization

The band

A G/wi band is a community (see Murdock, 1949:79) occupying a defined territory and controlling the exploitation of the resources of that territory. A discernible collective perception of unity and common purpose is apparent in the intraterritorial migration from one camp to another, in the conduct of camp affairs, and in the management of some aspects of interband matters. The stake in the common estate, which membership confers, is distinctive. The geographical separation of band territories imposes a further measure of distinctiveness on each band.

Although G/wi bands are discrete and have to restrict their sizes to accommodate their territorial resources, they are not closed communities in the sense that recruitment into them is confined to certain categories of persons. As I have indicated, membership is open to non-G/wi. Some bands even include non-Bushmen (Kgalagari Bantu) who have adopted the G/wi life-style and acquired conversational command of the language.

The band is the largest social unit within which sustained political and economic interaction occurs. The smallest among those investigated had an average membership of 25 and the largest band averaged 85 members.

Band formation

The formation of a new band is a comparatively rare event and no opportunity presented itself during the survey period for observation of the entire process. What happened in ≠xade band between 1963 and 1966 was recognizable as the initial stage of hiving off and confirmed informants' descriptions of what occurs when a new band is formed.

New bands are formed in response to overpopulation of a territory and an overloading of its resources or after the decimation and fragmentation of a number of bands by drought or disease. (Such catastrophes appear to have been fairly common in G/wi history; see Chapter 6.) In either case, a man gathers a nucleus of friends and kin together and they move into a suitable and vacant territory. When formation is prompted by overpopu-

138

The band migrates to a new campsite, each household carrying its possessions.

lation, hiving off extends over several years. The initial search for new territory is made in late summer. At this time travel is easiest, there is much visiting between bands, and the flow of news is at its height. Guided by the interband intelligence network, the founder leads a party of men on an extended hunting trip into country offering the most likely prospect of sufficient resources. Later they take their families with them and spend a month or more away from the parent band, returning before the annual breakup in winter. The following summer, once the wet season is established, the pioneers make their first visit and continue to spend longer and longer periods in the new territory. The composition of the group changes to some extent as some withdraw and elect to remain with the parent band or to move to other established bands, and their places are taken by others. As the absences of the pioneers in their territory grow longer and more frequent, the separate identity of their group emerges until, eventually, it is recognized as an autonomous band.

The G/wi keep no tally of the years and can seldom fix the past with any accuracy beyond about three years. There are few historic happenings, but the widespread smallpox epidemic of 1950 did provide a rare reference point that could be fixed from government records. Two bands, those centered on Easter Pan and G!ō:sa, were formed from survivors of other stricken bands, which were greatly reduced in number by the epidemic. I could not form any clear idea of how long the other bands had been established. The largest and the only band that showed and felt an excessive pressure of population on its territorial resources, ≠xade band, had apparently been established for at least 30 years and perhaps for much longer than that.

Band composition

The composition of the six bands that were closely analyzed tended to confirm the foregoing description of band formation and the implication that there is no clear structural factor that determines band membership.

Although informants were not unanimous, the majority favored marriage outside one's own band, residence with the wife's parents until after the birth of the first child, and then permanent residence in the band of the husband's parents. This was considered the ideal pattern. I found, however, that many couples had been members of the same band before their marriage and that the frequency of nonobservance of the ideal of band exogamy was proportional to the size of the band. The stated rule of ultimate virilocality – of moving to the band of the husband's parents – was followed to the extent of an excuse always being furnished for its breach.

There is a weak and imprecise notion of bilateral descent conferring optional rights of band membership, that is, one may claim membership in the band of birth of either parent. Obviously, descent does not predict which band one shall join. Analysis of band composition and of individuals' accounts of their histories of membership leads me to conclude that this descent "rule," in the virtual absence of lineages, is simply an expression of the fact that band membership is potentially open to all but that one is likely to be drawn to, and accepted into, the band whose company is most congenial to one's own preferences and personality, with the number and strength of kinship bonds exerting more influence on this preference than other factors.

"Ownership" and territorial rights

The only hint of the lineage concept is contained in the vernacular model of band "ownership." The founder of a band is the *! ū:ma* (owner), or the *! ū:sa* (feminine), of the territory and is said to be the one from whom visitors and prospective recruits to the band ask permission "to drink water" (i.e., camp with the band and share in the use of their territorial resources) and of whom recruits seek approval to join the band. This role devolves on the "owner's" eldest living descendant. In practice there is seldom a single "owner" but two or three and even four. Their role is more nearly that of spokesmen for the band; visitors and other newcomers, or their sponsors from within the band, asked one of the "owners" to be allowed "to drink your water" (even when the waterholes had been dry for months and would not fill again for more months). As advance news of movements in the territories of neighboring and allied bands is usually available, new arrivals are seldom a surprise and band consensus has already crystallized during informal discussion of the prospect of the newcomer's arrival. The "owner" then merely voices his approval on behalf of his fellows. In the rare instances of unexpected or unwelcome arrivals, an "owner" resorts to various delaying stratagems. It is on these occasions that the nature of the "owner's" role as spokesman, rather than as leader, emerges. He or she refers the matter to the other members of the band, sometimes without attempting to influence them one way or another. Although membership is not closed, it does confer exclusive rights. Permission is never actually withheld and its asking is simply a formality. It is, however, a formality that clearly indicates that the use of territorial resources and residence have to be granted before they are gained. Unwelcome visitors are given permission to remain but are later eased out of the band.

The suggested heritability of "ownership" is contradicted by the fact that "owners" of five of the six bands investigated were not demonstrably descended from the founders. In the sixth band the "owner" was the founder himself. Informants' opinions of who should succeed the then "owners" made it clear that no lineage principle was at work. "Ownership" would devolve, it appeared, on those older (but not elderly) members of long-standing who had forceful personalities. They qualify by their thorough knowledge of the band, its territory, its members and resources, and their established trustworthiness and ability in dealing with people. Long residence in a band inevitably establishes real and classificatory kinship links, and there is, therefore, a high probability of some kinship bond existing between "owner" and successor, but not necessarily one of descent.

Interband migration

An intermittent exchange of members occurs between bands. Whole households migrate to other bands for a variety of reasons. Many children marry into bands other than those of their birth, and divorced or widowed spouses move back into their old bands or away from unhappy situations. The gross annual rate of short-term migration is very high, exceeding 200 percent of the membership of some bands in years of good rainfall when there is plenty of food and travel is relatively easy. Determination of the rate of permanent migration, as distinct from visits of shorter than one year's duration, presented much the same difficulties as did census taking. Reliable data were gathered in only four of the six bands investigated. Between 1959 and 1966:

> ≠xade band: 5 households left and 3 joined
> G!ō:sa band: 2 households joined
> Easter Pan band: 3 households left and 2 joined
> Tsxobe band: 6 households left and 6 joined
> Piper Pans band: 6 households left, 2 joined (1960–1964)

The G/wi kinship system

The G/wi kinship system is characterized by a dichotomy of kindred (excluding spouses) into joking and avoidance/relationships, a virtual absence of lineage structures or other corporate kin groups beyond the nuclear family, and a ready facilitation of fission and fusion within and between groups of kin. From an ego focus, its primary extent encompasses three ascending and three descending generations in the vertical dimension. Horizontally it covers parents' siblings and their spouses and two generations of their descendants and the spouses of these descend-

ants. The system is bilaterally symmetrical in that there is neither terminological nor normative discrimination between matrikin and patrikin.
The scope of the system is widened beyond these categories by the institution of fictional siblinghood and the universalistic device of incorporating kin-of-kin into the network of joking and avoidance/respect relationships.

The figures illustrate the terminology of the system in its primary extent.

Joking and respect

The relationships an individual has with his kin are classified as either
joking or avoidance/respect relationships. The structure is ego-centered
and therefore varies from one individual to another, precluding a group-
wide uniformity in the dichotomy that could constitute anything resembling a moiety organization. In general, all kin are in a joking relationship
except one's parents and opposite-sex siblings and their classificatory and
fictional equivalents, who are in the avoidance/respect category. The relationships are dyadic and reciprocal, conceived of as existing between
pairs of individuals, each having the same category of relationship with
the other.

An avoidance/respect relationship, (*gjiukxekxu*) requires that those so
related should ≠*ao* (*v.t.*, to be reserved or respectful toward, to be scared
of) one another. Their proper behavior is characterized by:

> Not sitting close together, and generally avoiding bodily contact if not of
> the same sex
> Being careful not to swear or make bawdy remarks in the obvious hear-
> ing of those in an avoidance relationship
> Not touching their possessions without permission; if an object is to be
> passed between avoidance relatives, an intermediary should, properly,
> be used and a direct transfer avoided
> Younger persons use the honorific plural form when addressing their
> elders

Children are not expected to observe this etiquette rigidly until they are
7 or 8 years old. Within the confines of respectful etiquette, a great affection is displayed and expressed between parents and children and between brothers and sisters. The reserved behavior is consistent with the
reinforcement of the prohibition of incest between brother and sister and
with the exercise of parental authority over children. The latter is the only
G/wi relationship in which authority is intrinsic.

Although I have formally dichotomized the two types of relationship,
there is a gradation of decreasing reserve in joking relationships. Behav-

ior toward one's grandparents lacks many of the restraints of relationships with one's parents but is less free than that between cross-cousins of the same sex and approximate age. The last is ideally close and friendly. Property may be used without asking permission, conversation is unrestrained, and physical contact is the normal expression of friendship between young adolescent girls. Sharply critical comment on joking partners is permitted in public and conversational duels are frequent. Older boys and girls and, less commonly, adult men and women spice their exchanges with ingeniously bawdy gibes.

A joking or an avoidance/respect relationship is tied to each kinship term (except husband and wife). The terms in a joking relationship are:

> *babama* (grandfather)
> *mamasa* (grandmother)
> *nǁodima* (grandson, male cross-cousin)
> *nǁodisa* (granddaughter, female cross-cousin)
> *g≠wa'usa* (sister-in-law, potential wife)
> *gjiba:xuma* (younger brother)
> *gija:xuma* (elder brother)

For avoidance/respect relationships the terms are:

> *ba:ma* (father)
> *gje:sa* (mother)
> *ba:gǀwama* (little father)
> *gje:gǀwasa* (little mother)
> *gjiba:xusa* (younger sister)
> *gija:xusa* (elder sister)
> *gǀwāma* (son)
> *gǀwāsa* (daughter)

(This is appropriate to a male ego; in the case of a female ego, the joking/respect statuses of siblings are transposed and their term *g≠wa'usa* is replaced by its masculine equivalent, *g≠wa'uma* [brother-in-law, potential husband]. The translations include classificatory as well as real categories. The terms "little father" and "little mother" are literal translations of those terms used for, respectively, real and classificatory father's brother and mother's sister's husband, and mother's sister, father's brother's wife.)

In categorizing kin into joking or avoidance/respect relationships, symmetry is maintained so that (with the exceptions discussed below) any individual has the same type of relationship with a third person as does his joking partner. In a series of triads, the triangles are congruent (Figure 8); joking partners A and B also have a joking relationship with C. A congruent triangle of joking relationships similarly exists between D, E, and F. The avoidance/respect relationships between A and D is matched by

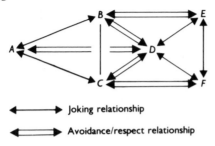

Joking relationship

Avoidance/respect relationship

Figure 8. Joking and avoidance/respect relationships among six persons, designated *A* to *F*.

similar relationships between *A*'s joking partners and those of *D*. This principle of congruency is coupled with an alternation of generations, for example, the parent-child avoidance that exists between Ego and his father and between his father and *his* father contrasts with the joking relationship between Ego and his paternal grandfather (Figures 9–12). In turn, Ego is in an avoidance relationship with his great-grandfather, should the old man still be alive.

The congruency principle also operates in the extension of the relationship network through a chain of "kin-of-kin." An individual may assume a quasi-kinship relationship with a hitherto unrelated person who is kin

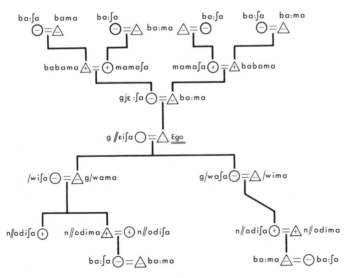

Figure 9. Joking (+) and avoidance/respect (−) relationships between Ego and Ego's ancestors and descendants. (ʃ = s.)

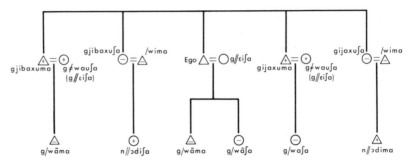

Figure 10. Joking (+) and avoidance/respect (−) relationships between Ego and Ego's siblings and offspring. (ʃ = s; ɔ = o.)

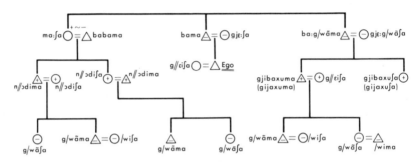

Figure 11. Joking (+) and avoidance/respect (−) relationships between Ego and Ego's patrilineal kin. *Gijaxu-* denotes older sibling; *gjibaxu-* denotes younger sibling. (ʃ = s; ɔ = o.)

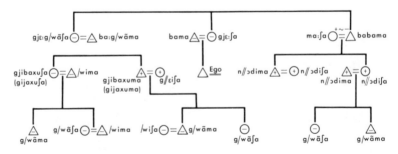

Figure 12. Joking (+) and avoidance/respect (−) relationships between Ego and Ego's matrilineal kin. *Gijaxu-* denotes older sibling; *gjibaxu-* denotes younger sibling. (ʃ = s; ɔ = o.)

to a kinsman. The joking/respect roles in the triad are determined by the respective kinship links, in keeping with term-tied statuses, and work out as congruent triangles.

The principles of congruency and of alternating generations are both conditionally negated by the fact that Ego has joking relationships both with his cross-cousins and with their parents. The G/wi explanation of this situation is ingenious and throws some light on their perception of their own social structures. In the vernacular model the ideals of band exogamy and virilocality combine to locate cross-cousins in bands other than one's own. Interaction with them is, therefore, infrequent, and the problem of incongruent triangles seldom arises. Cross-cousins of opposite sex are preferred marriage partners. Clearly one cannot marry all of them, but the parents of the one taken as a wife immediately cease to be joking relatives, *babama* and *mamasa*, and become *ba:/wima* (father-in-law) and *gje:/wisa* (mother-in-law), who are in the avoidance category. The remaining cross-cousins, according to the G/wi explanation, will be separated from their parents by marriage, ruling out incongruous triads. There are, therefore, two mutually exclusive forms of interaction possible, either of which solves the problem of awkwardness that is inherent in incongruent triangles.

However, endogamy is common in large bands and cross-cousins are, in fact, brought together in such bands. Informants did not verbalize the cognitive calculus that solved this problem, but the practice is for those cross-cousins who are impossible or unlikely marriage partners (being already married or of inappropriate ages) to treat one another as siblings and their respective parents as "little" (or classificatory) parents. The transformation is not carried through to the generation of the parents themselves, so their own joking/avoidance balance is not disturbed by the change. Such ad hoc distortion of formal structure is typical of G/wi social organization and will be discussed later at greater length.

A system of categorization that selects a large proportion of a universe for inclusion has two possible organizational strategies. First, it may contain a large enough number of specialized, narrow categories, each with distinctive criteria of qualification, to be able to accommodate every possible member selected from the population universe. Alternatively, the system may have a limited number of specialist categories but must, then, have one or more broad categories into which can be placed any residual members not qualified for specialist classification. The G/wi kinship system, in its extension of kindreds, is based on the second strategy, having a few narrow, specialized categories and a small number of broad, unspecialized categories. In contrast to the numerous categories of the first strategy, in which qualification criteria are necessarily exhaustive,

Table 7. *Breadth of kinship term categories*

Kinship term	Person to whom term may be applied
ba:ma	F, MH (the two are distinct in instances of remarriage after divorce or widowing)
gje:sa	m, fw (the two are distinct in cases of polygyny or remarriage)
ba:glwāma	FB, FFBS, FFFBSS, MFBS, MMFBS, MFFBSS, H, of *gje:glwasa*
gje:glwāsa	mz, mmzd, mmmzdd, fmzd, ffmzdd, fmmzdd, w, of any *ba:glwama*
babama	MB, MMBS, MMMBSS, FMBS, FFMBSS, FMMBSS, H, of any *ma:sa*
ma:sa.	fz, ffzd, fffzdd, mfzd, mmfzdd, mmffzdd, w, of any *babama*
gjibaxuma, gijaxuma	B, FBS, MZS, FFBSS, MMZSS, FWS
gjibaxusa, gijaxusa	z, mzd, fbd, ffbdd, mmzdd, mhd
nllodima	SS, DS, ZS, MZDS, FBDS, son of any *babama / ma:sa*, H of any *nllodisa*
nllodisa	dd, sd, zd, mzdd, ffbdd, d, of any *babama / ma:sa*, w of any *nllodima*
glwāma	S, DH, BS, "BS," WS, HS, S, of any *nllodima / nllodisa*
glwāsa	d, sw, bd, "bd," wd, hd, d, of any *nllodima / nllodisa*

the criteria for qualification in the broad categories are few and tend to be nondistinctive. A large measure of versatility is permitted. Individuals may vary their categorization with changes of the social context and field of interaction, and in the same category one member may readily be substituted for another. Table 7 illustrates the breadth of the terminological categories.

Terminological equivalence is matched by an approach to role equivalence. The parent-child link contains the most specialized category, and role equivalence between natural parent and stepparent is only complete where the stepparent assumed the role early in the life of the child but is less than complete in respect of *gje:dzi* (mothers) in polygynous households. The role equivalence of the reciprocal term *glwā-* (child) is approximated by stepchildren or "half-children," that is, where the term is applied to members of the household. The roles of all *glwāna* (children) outside the household are equivalent among themselves. As mentioned later in the section dealing with marriage, classificatory siblings may become half-siblings. If this occurs early in the life of a child, the distinction between real and classificatory siblinghood disappears and the role equivalence of the two categories is complete. Beyond the household, the roles of all classificatory siblings are equivalent.

Terminological equivalence thus reflects extrahousehold role equivalence, and outside the household the kinship system facilitates substitution of any one individual in a category by another in the same category. This ease of substitution outside the boundaries of the household reflects the independence and solidarity of the domestic group itself. From this independence, and from the facility of substitution among persons included in the extrahousehold kin network, stems the ease with which the processes of fusion and fission can occur, both within the band and between bands. The combination of low fission inertia (i.e., low resistance to processes of separation) and high fusion valency (i.e., a readiness to join together) are features of major importance in the socioecosystem of the G/wi.

Kinship and marriage

A G/wi girl is married when aged between 7 and 9 years to a boy some 7 years older. My informants stated that the preferred marriage is one between cross-cousins, between a boy and the daughter of either his real or classificatory mother's brother or father's sister. The couple should live attached to the household of the wife's parents until their first, or even second, child is born, after which they are regarded as an autonomous household and should go to live among the husband's people. The view of informants was that brother and sister would not normally spend their adult lives in the same band, which, together with the preference for cross-cousin marriages, implies a strain toward band exogamy.

As I have already mentioned, actual behavior does not always accord with what people told me ought to be done. My analysis of existing marriages and band members' speculation about the future spouses of single boys and girls revealed a marked discrepancy between the evident vernacular ideal of cross-cousin, virilocal, band-exogamous marriages and actual practice. Of the 73 marriages adequately analyzed during the survey, only three were between men and the daughters of their father's sister and five married daughters of their mother's brother, that is, only 11 percent were cross-cousin marriages. My attempts to elicit explanations of the divergence between preference and practice were fruitless, and it became clear that I was asking questions based on my literal interpretation of the stated preferences and that these questions were meaningless to the informants. What did emerge was that all first marriages (as distinct from subsequent marriages of widowed or divorced persons) had been between boys and girls in a *n‖odima-n‖odisa* relationship, which is the

149

terminological equivalent of cross-cousinship and, hence, placed the two in a joking relationship. It was not possible to obtain details of the relationship that had existed between the respective parents of couples before the latter had married. However, the speculation about the future marriages of boys and girls did throw a good deal of light on this matter. Approval of a suitor is much readier if he and his parents are well known to, and on friendly terms with, the girl's parents. Answers to further questions and my own observations confirmed this indication. This suggested another interpretation of the marriage "rule," namely, that the kinship model of cross-cousinship was being used to express the ideal prerequisites of marriage. The kinship model contains three elements: First, the boy and girl should be joking partners and not in an avoidance relationship; second, the two sets of parents should be on friendly terms and know enough about each other's children to judge their mutual compatability of temperament; and third, the parents should have the trust and affection of the couple and thus be able to guide them in times of marital stress. This last aspect is of particular relevance to the girl's parents during the years of initial uxorilocality, the period of residence with the wife's parents. An additional feature of the kinship model precludes intermarriage between the two sets of parents should death or divorce end their marriages and leave one of each set single and eligible for remarriage. From the point of view of the young couple, marriage between a parent and a parent-in-law would result in an intolerable rearrangement of the statuses of the spouses and their mutual kin.

Support is lent to the model-using hypothesis by the commonly held attitude that boys and girls choose their own spouses and are subject to parental approval, rather than to parental direction. Uninhibited gossip and speculation must have some influence on boys and girls in their choice of wives and husbands, but no formal pressures are exerted on them to direct their choice. Overt courting is preferably initiated by the boy, and even though precocious, outspoken, and wilful girls are permitted a good deal of latitude in dealing with shy suitors, it is the boy who should seek permission for the marriage from the girl's parents. Her parents' proper behavior is to be initially defensive and protective, and however much they may actually concur with prevailing opinion favoring the marriage, they should rebuff the boy's first advances. Typically, they claim that their daughter is too young and foolish to know her own mind on a matter as important as marriage. Invitations to occasional meals and other contrived situations serve to keep the suitor on a string if he is favored by the girl and her parents. If they do not approve of him, the parents will be quite blunt and tell him so. Their veto can be eroded by persuasion if the couple are determined to marry and have the support of a substantial number of band members. If the parents' objections are backed by public opinion, it becomes everybody's business to keep the couple apart, and the young man is eventually forced to abandon his suit.

There is no ceremony marking the commencement of married life; the couple simply build themselves a shelter or hut adjacent to that of the girl's parents. They address one another as *kxaoma* (man, i.e., husband) and *g//eisa* (woman, i.e., wife) and adopt the appropriate affinal terminology in respect to one another's kin. According to my informants, a boy would treat his *babama* (MB, FZH) and his *ma:sa* (fz, mbw) as avoidance kin as soon as he began to court their daughter, anticipating their conversion from joking kin to avoidance affines by the fact of his marrying their daughter.

A girl's menarchial ceremony (see Silberbauer, 1963) marks her transition from childhood to womanhood and illustrates a number of G/wi values and attitudes. Parent-child bonds are emphasized by the mother's role in guarding and attending her daughter during the latter's seclusion prior to the ceremony. The mother also shepherds her daughter through the early stages of the ceremony. That a husband is considered to be weakened and abnormally vulnerable to danger when his wife menstruates indicates the strength of the bond between them. He will not hunt, nor even touch his weapons, lest an accident befall him. When a girl goes into menarchial seclusion, her husband leaves the conjugal shelter or hut and retires to the bachelor's shelter, where he is cared for by the men of the band until his wife's menstrual flow has ceased and the menarchial ceremonies commence. He joins his wife for the last and longest phase of the ceremonies. The couple are tonsured and their bodies are tattooed in matching patterns of lines and inoculated with the blood of one another's wounds. Magical herbal preparations are mixed with this blood to ensure a harmonious married life and to ward off want and disaster from the couple. While seated together, husband and wife are individually admonished to observe the conjugal virtues of patience, discretion, loyalty, and affection. There is no mention of fertility and the focus of this part of the ceremony is on the relationships between husband and wife and with other band members. It is clear from this affirmation of marriage, following the rite of passage into womanhood, that a marriage, although intended to be lasting, is likely to be threatened by social as well as environmental forces. It is also apparent that chastity in either partner is a matter of demeanor and discretion rather than of actual sexual behavior. The repeated reference to the couple's mutual dependence emphasizes the complementarity of their roles and denies any view of institutionalized dominance of either partner.

When the tattooing is completed, the girl's mother takes her by the hand and gently leads her out of the seclusion hut. She holds to her daughter's forehead one after another of the food plants available at the time, pronounces the name of the plant and tells her daughter its uses and virtues. Then, stretching her arm out and pointing around the horizon, she introduces the band territory to her daughter. "This is the country of all of us, and of you; you will always find food here" (which also has the connotation of "you will always be at home here"). The girls and young women in attendance then run the girl through a symbolic shower of rain, which, apart from being intended to ensure that she does not suffer drought in her life, associates her with the life-force of rain. This is not symbolic of fertility as such but of a wider view of life, which includes survival. Informants explained that the association would make the girl beloved of people among whom she would live.

The girl and her husband are then decorated in matching patterns of ocher stripes and the men of the band begin to gather near the seclusion hut. The girl's father comes forward, takes his daughter by the hand and gently leads her (she is supposed to be blind at this stage) to her husband's hunting weapons, which are placed against a tree amid the watching crowd. The father guides his daughter's hand to the weapons and introduces them to her "so that they will not harm your man nor anybody else" (i.e., so that they will never be used in anger or be the cause of accidents). Then, turning his daughter to face the onlookers, the father plucks a handful of grass, forms it into a small bundle, and holds it before the girl's eyes. He then snaps it in two, so restoring her eyesight. The first of the bystanders is then introduced, "See your people. Here is So-and-so . . . [Then follows a who's-who sketch.]. See him [or her] wherever you go."

The complementarity of the parents' roles in their daughter's ritual translation into womanhood is consistent with the values expressed to her and her husband during the ceremony. Understandably, it is the mother who cares for the girl while she is actually menstruating; contact with her would be dangerous to men at this stage. The mother is present when the conjugal virtues are impressed on the couple and when the magical precautions against troubles are applied and it is she who introduces her daughter to the band's territory and its resources. The father's part lies in safely reincorporating the girl into the male world of hunting weapons and into the social context of the band as a whole and its individual members. (This view of the parity of parental roles is contrary to that expressed in my 1963 paper; I have since amended this earlier assessment in the light of subsequent observations.)

The uxorilocal phase of marriage

There appears to be a period of adolescent sterility among G/wi girls, who do not usually conceive until they are 16 or 17 years old. The uxorilocal phase of marriage, which lasts until the birth of the couple's first child, may be, therefore, as long as 10 years. It is a period of apprenticeship for both spouses; although the girl can handle household tasks by the time she is married, she continues to gather foods and to run her household under the eye of her mother. From the gossip and conversation of the women with whom she and her mother spend the day, she furthers her education in personal matters and those pertaining to the band. The husband, under the tutelage of his father-in-law, perfects his hunting and other subsistence techniques and learns from the older man and his cronies the conduct of band affairs. The husband occasionally hunts with his father-in-law when the band is gathered in one camp and during the time that the extended household is in isolation during winter and early summer. He makes a present of part of every game animal he kills but not birds and small mammals, to his parents-in-law. (Plant foods are not shared unless drought is severe and food is short.) As the relationship between the young husband and his parents-in-law is usually one of affection combined with respect, the parents are able to influence the couple's reactions when conjugal tensions arise. That such tensions seldom develop into serious threats to the safety of the marriage is due partly to parental influence and partly to the fact that the young wife and husband are seldom thrown onto their own unaided resources; only when in the doubtful privacy of their shelter, standing next to that of the parents, are they likely to be alone together. For the rest, their field of interaction includes the whole household, the two couples and the wife's young siblings.

However, should the safeguards prove inadequate and conflict develop to an intolerable level, the marriage will be recognized as a failure. The husband leaves, either to return to his band or to the bachelors' tree in the middle of the band camp, and the girl simply moves back into her parents' shelter. The consequences of the marriage may take some time to disappear; the use of affinal kin terms and the associated behavior patterns may persist for a while. If the breakup of the marriage is accompanied by schism within the band, a result of members taking sides in the conflict (which is more common in band-endogamous marriages than in exogamous marriages), the household of the wife or the husband or both may migrate to other

153

bands. Generally, however, a broken "apprenticeship" marriage is regarded as a regrettable, but inevitable, result of incompatible personalities and gradually fades from public thought.

The termination of the uxorilocal phase

The wife's first confinement, in which she is attended by her mother and older women of the band, usually marks the end of the couple's uxorilocal dependence. When the band next moves its camp, the couple build their shelter away from that of the wife's parents. If the season is one in which travel is easy, they might migrate to another band after a few months. As I have said, the correct choice of new residence is considered to be the band in which the husband's parents live. However, this stated preference is only followed to any extent in cases of band-endogamous marriages, that is, those in which the couple have simply remained in the joint band of origin. The proportion of autonomous couples remaining in the bands of their parents was, even then, only about half. All couples furnished some explanation for not having moved to the husband's band; they had postponed the return and were only "visiting" (albeit for several years). Men married to only or youngest daughters pointed out that their parents-in-law were "alone" and needed their daughter's company. Some wives flatly refused to leave their parents; men who had married into small bands said that there were few people and the band needed more members. These explanations appeared to be acceptable, for I was not aware of any pressure being put on these couples to move to the "proper" bands. Where a man has taken his family back to his parents' band, the frequency of interaction between the two households is seldom markedly higher than that between other households. When first gathering census and genealogical data, I was sometimes surprised to discover that a man and his parents were present in the same band; the parents would be identified to me as a couple with whom I had hardly ever seen the man speak.

There is the problem of why the G/wi should profess an ideal when, in the first place, only a small minority of them actually follow it and, second, when it constitutes an ecological luxury that they cannot afford. Some light was shed by the evidence of those households that had moved out of the band of the husband's parents after having been virilocal for some years. Informants felt that such a move was quite in order and that it called for no excuses or explanation. It does seem that the ideal only calls for a period of virilocality

154

and does not demand that it follow immediately on the uxorilocal, initial phase of the marriage. This interpretation is consistent with the actual composition of bands and with the histories of household migrations. If this interpretation is correct (and my informants agreed with it), the position is that there is a period of uxorilocality followed by a neolocal choice of residence that ought to include some time spent with the band of the husband's parents.

Polygamy

Polygyny is permitted but is not common, only 9 husbands having more than one wife. There was only one instance of polyandry. A woman had left her husband for his widowed best friend. The two men "fought" and the second husband went away with the woman. However, a triangle has three sides; and the two men missed each other, and the wife missed her ex-husband, who, in turn, had refused to take a second wife because he missed his first wife so much. The absconding couple returned after two years and the original husband moved in with them. The band had discussed this solution to the problem at great length before the return of the couple. Although the polyandrous solution was unprecedented, it was accepted by everybody. Band consensus had devised the solution and friends persuaded the three concerned to try it out. The second husband and the wife occasionally went away on visits, leaving the first husband behind. He and they appeared to be content with the arrangement, which was still in being when I left the field late in 1966.

Of the polygynous husbands, one had four wives and eight had two. Three husbands had each taken as a second wife the younger sister of his first wife because the sisters had not wanted to be parted. Only one of these husbands had remained with the band of his parents-in-law after the obligatory period of uxorilocality, all other polygynous households having taken up neolocal residence in bands containing one or more siblings of one wife. Five of the husbands had married widowed or divorced mothers of young children who were dependent on them at the time of the second marriage. The four wives of the one man were neither siblings nor were theirs second marriages; they had all apparently been anxious to marry him. All but one of the polygynous husbands appeared to be rather disappointed with their lot and privately expressed the view that polygyny was a burdensome and hazardous form of marriage because of the threat of disruptive jealousy among the co-wives. The attitude of monogamous men was that polygyny was a good enough solution to the problem of lonely sisters and divorced or widowed mothers of young children.

Co-wives occupy separate but adjacent huts or shelters. Although each gathers and prepares food for her own children and for herself, the co-wives usually travel together when out gathering and assist each other in lifting and carrying loads. They also help one another in hut building and other domestic tasks, including the care of one another's children, to whom tidbits, but not regular meals, are fed.

Implicit in polygynous households is the problem of sexual relations. A woman must not conceive until she has weaned her youngest child, normally when the child is 3 or 4 years old. Informants were unanimous that abstinence is the only means of contraception and that this was a grueling test of a husband's willpower, as the responsibility for avoiding conception was his. I did not inquire into the stresses that might be imposed on a polygynous household by the sexual accessibility of only one of the co-wives, nor why the greater opportunity for intercourse did not enhance the rather low popularity of polygyny as a form of marriage. These were rather delicate topics.

Divorce

I have already mentioned that the marriage of a young, childless couple may be ended without formality. All marriages may be dissolved by unilateral decision of either spouse or by mutual consent. The only semblance of formality is the act of either partner's leaving the conjugal hut and taking up residence elsewhere after having declared his or her intention not to return. Incompatibility of temperament and flagrant adultery are the recognized causes of marital breakup. Incompatibility is revealed by frequent arguments between the couple and accusations of laziness, stinginess, and *xamxasi* (literally, lion-like, i.e., brutality, violence, and excessively short temper), which characteristics are regarded as serious character defects by the G/wi. Any threat to marriage is a matter of interest to the band. Apart from the excitement that the drama of domestic strife may lend, there is genuine concern that conflict not develop too far. Friends and close kin readily intercede to avoid a breakup. An adulterous spouse is quietly told by friends to mend his or her ways, or if it can be done without precipitating marital disaster, the offending husband or wife is publicly teased by his or her joking partners. De-

156

meanor is the important factor; even where the inference of adultery is inescapable (e.g., when a wife returns 4 months' pregnant after a 6-month visit to another band), no objection is raised by the husband or anybody else. Should, however, the adulterous genitor lay claim to the child when it is born, or should he or the wife boast of the liaison, a fight and possibly divorce will follow. Even where liaisons were open secrets in the band, they were tolerated if no overt indications were given that would force the husband or wife to take public cognizance of the partner's infidelity.

The "fighting" that accompanies accusations of adultery by either spouse consists of the exchange of a few blows and a great deal of shouting and swearing. The aggrieved spouse is allowed to carry on for two or three days and is then quieted by the persuasion of the rest of the band to accept the situation.

The rate of divorce could not be determined because of the virtual impossibility of establishing any sort of chronology. Furthermore, the bands are anxious to forget past conflicts. Among the "ever-married" individuals investigated, only 3 percent were recorded as having been divorced. All but two of these had histories of only one divorce. The two exceptions were women of some notoriety who wandered from one band to the next collecting a new husband in each and then leaving him and the band after one or two years. They were cheerful and vivacious women who could not avoid discovery of their many adulterous affairs. (Their reputations probably alerted so many women and intrigued so many men as to make discovery inevitable.)

The data I gathered on divorce relate only to those divorces that occurred during my stay or happened shortly before my arrival in the field. The data are obviously too conservative as no account is taken of earlier divorces.

Custody of children after divorce

Both men and women firmly stated that the custom is for the children of a divorced couple to remain with their father; a child still at the breast stays with its mother until weaned and then goes to the father. In the small number of divorces I was able to investigate, this was indeed the situation. The G/wi themselves were unable to explain the rule of paternal custody beyond saying that "children belong to their father." Children visit their mother if she has not moved to a distant band, but both the children and the mother feel the separation deeply. If both spouses remain in the band after their divorce, the children spend a good deal of time with their mother but clearly regard the father's hut or shelter as "home."

The threat of being deprived of her children must deter a mother from risking divorce, but it is difficult to see any other explanation of the custody rule. It raises the suspicion of the existence of a patrilineage, with which paternal custody would logically be associated. However, there is no heritable status other than the freely available right of band membership, and there is no other evidence of a lineage principle in the kinship system. The importance of the mother's role is at least equal to that of the father in the rearing and socializing of the children, and the strength of the emotional bonds between mother and child is no less than that between father and child. The eligibility of the spouses for remarriage after divorce is equal and no explanation of the dominance of the father's rights to the children can be found in the respective economic roles of the parents; it is the mother who gathers and prepares the greater portion of the household food. I can only remark that this is one of the few instances in which G/wi men have a status superior to that of their women.

Remarriage

Elderly widows and widowers usually remain single after the death of their spouses, but younger widows and widowers and divorcees are said to remarry as soon as possible after separation. How long this may take is determined by the availability of eligible spouses. As I have indicated, polygamy is not very common, and a widower generally has to find an unattached widow or divorcee and vice versa. Excepting the marriages of couples whose adulterous liaisons occasioned their prior divorces, second marriages appear to be more arrangements of convenience than love matches. The range of choice of second spouse is small, being restricted by the low incidence of death and divorce among middle-aged and younger couples. The institution of avoidance relationships precluding marriage further narrows the choice. However, despite their lack of choice in selecting a second spouse, second-marriage couples seem to settle down as well and to form as happy households as do first-marriage couples.

The guarantee that the kinship system provides for widows and divorcees in the form of a claim on their sisters' husbands or their former husbands' brothers as husbands is commonly referred to but seldom acted upon, and I did not encounter any such marriages. Informants told me of two sororal and five leviral marriages of widows with dependent children, but these involved people in bands that I did not visit and I was not able to investigate these marriages personally.

The household

The household is the most stable and enduring group in G/wi society. It does not divide when the band splits up in winter and is the normal unit of interband migration and of the formation of new bands. A household is the unit of consumption of gathered plant foods and water brought into camp, of building materials, and of small game and birds. It furnishes its own labor force to obtain and prepare these commodities. In its simple form, a household consists of a married couple and their unmarried daughters and sons below the age of puberty. (Older, unmarried sons move out of the family shelter to the bachelors' shelter in the middle of the band encampment and return to the family for occasional meals and when the winter separation occurs.) In extended form, the household includes a dependent daughter and son-in-law. A polygynous household is comprised of the several wives and their children, including dependent married daughters and sons-in-law. Some households have quasi extensions of one or even two elderly parents of one or both spouses. Although separate shelters are occupied by the two senior generations, food, firewood, and other commodities are shared to some extent and the whole unit migrates together between campsites, into winter isolation, and between bands.

The domestic cycle

An overall cyclic pattern of development occurs in households over their lifetime. The phases of the cycle do not fit neatly into Fortes's (1966) schema, but his terminology is followed where it is appropriate.

The initial phase. A household begins life with the marriage of a young couple. In its initial phase the new household is an adjunct to that of the wife's parents and is, therefore, uxorilocal. The households migrate together as one, building adjacent shelters and staying together during the period of isolation in winter and early summer. The daughter's daily routine is governed by her mother as the two gather and prepare their food together. Although they do not normally share their food, many of the domestic tasks are shared. The young husband is subject to the authority of his father-in-law and is his apprentice during winter isolation, helping with traplines and hunting under the older man's direction. During the seasons when the band lives together, the son-in-law spends more time with other young adults and hunts only occasionally with his father-in-law. A share of all but the smaller mammals killed must be given to the father-in-law. The latter also shares his meat but within a larger circle. If the daughter is the youngest in the family, her father may be past his best as a hunter and bring in no more, or even less, than the young and initially inexpert son-in-law.

As mentioned earlier, the period of uxorilocal residence generally lasts from 6 to 10 years, until the first child is born, by which time the young couple have perfected their subsistence techniques and are socially mature enough to set up their own household and participate in band affairs without the guidance of the wife's parents.

The expansion phase. After completing the obligatory period of uxorilocal residence, the neolocal household typically consists of husband, wife, and their infant child. From this stage onward, the household is fully autonomous. Even if the family should take up patrilocal residence, it is not obliged to submit to direction by the husband's parents nor to follow them into winter isolation and, in fact, does not normally do so.

Expansion and dispersion. Because of the comparatively late age of weaning, children are spaced at least 4 years apart. By the time the third child is born, the eldest, if a daughter, will normally have married. If the eldest is a boy, he will have moved to the bachelors' shelter, or if the gaps between children are longer than 4 years, he may well have married and left the parental shelter and perhaps the band for that of his wife's parents. The mother's load of child care does not normally extend beyond the full-time care of one infant and the part-time care of an older, but still dependent, child. Few mothers have more than four children and most have three or fewer, so a woman's years of bearing children may be over by the time she is 30 or may continue 4 or 5 years longer, depending on how the children are spaced. I did not see any infants' mothers of greater apparent age than about 36, which is a middle age under the trying conditions of the G/wi, who age early and appear to have a life expectancy of about 45.

The processes of bearing children and rearing them as dependents until nearly the age of puberty continues for 18 to 25 years, during which the size of the household increases as the children are born into it and married daughters bring in their husbands. Any quasi extensions of attached elderly parents are acquired during this penultimate phase in the history of the household.

Replacement. After some 25 years of autonomy the household will have shrunk to a strength of two; children will have left to form their own autonomous households and the couple's parents will have died. At an age of more than 40, the husband will be an old man with failing strength and sight, no longer fit for hunting large game with much hope of success. To an increasing extent, he must turn to such minor semispecialization as handcrafts and child minding. In return for these favors, he receives gifts of meat from younger hunters during the seasons when the band is camped together. In the reduced household with only two mouths to feed, the wife can manage her subsistence tasks in years of normal plenty but will be hard-pushed in drought years. It is at this stage that the elderly couple may turn to their children for support and attach themselves to the household of a daughter or son and follow them into winter isolation, when the reduced field of interaction and reciprocation of favors restricts the contribution that the old man's efforts make to the household diet. Among the G/wi, death follows the onset of senility by only a few months, so the period of semiparasitic dependence of the elderly is brief.

Household relationships

The conjugal dyad. Public display of affection between husband and wife is considered childish and appropriate only to young couples in their first years of marriage. Older autonomous couples make no apparent effort to spend free time together. Their conversation in company is impersonal, almost cool, and their greetings, even after fairly prolonged absence, are offhand and devoid of any show of pleasure. However, the bonds of affection are strong, and in private conversations with my wife and me many informants revealed intense emotional ties with their spouses.

The strength of the conjugal bond is also indicated by formal aspects of the social structure:

1. To some extent there is a merging of the separate social identities of the spouses on marriage. They assume a common residence and adopt one another's kin as affines. The husband comes under the tutelage of his wife's father in a dependent role, which approximates that of a child in the family. The joking/respect status of the husband within his wife's kin group is the same as hers and vice versa.

161

2. In the menarchial ceremonies, the obvious symbolizing and the overt emphasis of the unity of a married couple and, within that unity, their complementarity and interdependence are clear evidence that the conjugal dyad is regarded as a unique and irreducible relationship.

3. The spouses have mutual sexual rights, which are exclusive to the extent that the community will defend these rights against any person who, by overt act or indiscretion, implies a claim to share them. This is the only wrong for which an aggrieved person is permitted physical violence upon a nonmember of his or her household. In polygynous marriages, sexual rights are guarded not only against outsiders, but the equality of those rights is also jealously guarded by the co-wives.

4. Although common parenthood of children is possibly shared by successive spouses, and is approximated to by classificatory parents, it is a status unique to the extant conjugal dyad.

5. The husband's susceptibility to danger during his wife's menstrual periods is a further indication of the strength and closeness of their relationship.

6. Once the couple has attained independent status, the husband is vicariously liable for any malicious damage inflicted by his wife and is obliged to make reparation for damage occasioned by her carelessness. He is also liable for damage done by his children, including a dependent son-in-law. This illustrates the extent to which the identities of husband and wife are merged, as liability is transferred on marriage to the man's father-in-law for the duration of the initial phase of the domestic cycle.

7. An adult married man does not normally receive food from any woman other than his wife, and a married woman does not normally give food to any man other than her husband. In the games that children play in imitation of adults, the girl who plays at cooking for a boy is his wife. To use prepared food as a symbol of the conjugal dyad is, among hunters and gatherers to whom food is of such great importance, an indication of the weight attached to the relationship.

The conjugal dyad involves a unique relationship in which nonmembers of the couple's household do not share. Although the relationships that spouses have with nonmembers are important, they are charactrized by a large measure of substitutability and alternation of personnel and are thus not unique.

Parent-child relationships. This is the only relationship in which authority is inherent. Parental authority is exercised in terms of reason in an atmosphere of affection and kindness. All kin, in the G/wi ideal, *wiku* (love one another), but among all relationships, parental love for children is the most openly and frequently demonstrated by acts and words, the frequency and intensity of demonstration being inversely proportional to the child's age. Babies are not normally left alone. When asleep, they are either cradled in the arms of a parent or are carried in the mother's cloak (which is tied about her waist to form a sling) in contact with her body. When a baby wakes, he or she is played with, fondled, and kissed by anybody in the immediate vicinity. A baby is given the breast whenever he or she cries. Attention lessens when the child is able to walk with sufficient skill to avoid or overcome obstacles without stumbling and falling. (This is at about 1 year of age; G/wi babies do not go through a crawling stage and seemed to me to be somewhat precocious in their acquisition of motor skills.) Only then is the child allowed out of arm's reach of a responsible person. Affection is given by all members of the band, but the child drinks only at its mother's breast. At this age, an awareness of the father's identity could perhaps come only through the child's association of him with the mother and with the household as these are the only nonverbal contrasting features that would distinguish the father from among the men who shower affection on the child.

The socialization and training of the small child are group efforts in which neither parent is especially prominent. Direction is given by the nature of the responses to the child's actions and by encouragement to imitate. The teaching of walking, dancing, and clapping in time are accompanied by physical support and manipulation of the limbs. Group responses include a wide spectrum of expressions of approval and mild to moderate alarm (which is the negative response).

The post-toddler stage leads shortly into the verbal stage, in which the child is able to comprehend the full range of simple statements, instructions, and questions appropriate to communication with a 2½- to 3-year-old and is able to make known his or her wants, ask questions, and make simple statements of state and action. At this stage the child clearly knows the identity of all members of its household, to whom it refers by both the correct kin terms and individual names. Training and socialization are now divided between the household and the play group. The play group comprises children of up to about 6 years of age and occasionally includes others up to 10 years old. In the household the authority of the parents begins to emerge as instructions are given in an expectation of obedience. Disobedience is met with expressions of disapproval and very mild physical punishment. The G/wi attitude is that any more than symbolic punishment is inappropriate in the young; "if a child does not hear when you speak to him, he certainly won't hear when you beat him." Until he has developed some sense of responsibility, it is better to reason with a child or, if this does not help, to initiate some other activity to distract him or to remove the object of naughtiness. However, parents do lose patience, particularly with children who are approaching the age of responsibility (about 6), and resort to a sharp slap with the open hand and sharper words. Irritability is permitted a tired parent, but rage is considered shameful behavior.

Differences in the amount of authority wielded by either parent are determined by the balance of their personalities. As the exercise of authority is confined to the household, the lack of a structural distinction between paternal and maternal authority does not conflict with other aspects of the role-set of either parent.

Specific skills are taught by the parents, with appropriate distinction in terms of the sexual division of labor. The skills are carried into the play group and perfected by practice and by further learning from others in the group. An elderly man or woman in the band keeps an eye on the play group while the parents are out gathering and hunting, and the child minder also plays an important part in teaching and training.

Under the circumstances of little cultural change and almost no secrecy, children's subculture is not separated from that of their elders by any distinguishable break but expands directly into the adult world. The knowledge, training, and socialization of the child in the play group thoroughly reinforces that which is received in the household from parents.

The social identity of the child is distinct from the name giving onward. A baby is named by the grandparents or by one of the siblings of its parents in a public announcement. The name usually commemorates some happening or circumstance associated with the birth. From then on the baby is properly referred to or addressed by his or her name or kinship term and not as "the baby" or "the child" (the two words are the same in G/wi). When brought to a band for the first time or when brought back to its own band after an absence of more than a few weeks, the baby is welcomed by all the band members, in turn, who touch him on the upper lip and say "Halisima [or whatever the baby's name is], I am the woman !xai!xai. Remain happily with us."

Although he or she has individual identity, a child continues to be dependent until parental status is achieved. The child's dependent

164

status is demonstrated by the vicarious liability of the father for its acts and by the manner in which the parents give instructions in the expectation of obedience. Parents may expect obedience from their children for the rest of their lives, but it is appropriate etiquette for a parent to make requests of independent offspring and not to give orders to them.

Children's lifetime attitudes toward their parents were summed up by an adult informant talking of his father and mother: "I love them and I fear [respect] them as I do no others and will always obey them because they made me." Although parents are a source of comfort, food, affection, and security, they are in the avoidance/respect category of kin. From the time the child is considered a responsible being, the physical and social person of the parent is treated with great respect.

Sibling relationships. The gap of 4 or more years between births puts the elder child close to the age of assuming responsibility by the time the younger is born. By the time the baby has learned to walk, the elder sibling is old enough to be of considerable help in caring for him or her. When the stage is reached at which the child joins a play group, the older sibling will have a special responsibility, particularly when the household is in winter isolation. In the case of same-sex siblings, a good deal of training in techiques of subsistence is in the older brother's or sister's hands. In exercising responsibility, the elder sibling has no individual authority but vicariously invokes that of the parents by delegation, that is, falls back on them to validate his instructions. Consistent with this vicarious authority, the relationship between same-sex siblings before the younger attains independent status is a mixture of joking and respect. The personality of the elder is not an object of respect, but his or her instructions are obeyed. In return, the younger sibling, under the tutorship of the elder, depends on the latter for help, material items, and security. The typical relationship between same-sex siblings is one of affection and great trust. Relationships between siblings tend to be dyadic, as the eldest brother or sister will be married and out of the household by the time the third child is able to distinguish individual identities, leaving only the middle and youngest child together in the household while the last grows up. That is to say, there are usually much closer bonds between first and second and between second and third than there are between the first and third children.

The behavioral inhibitions imposed by the brother-sister respect/avoidance etiquette reduce the extent and frequency of interaction between opposite-sex siblings and place greater social distance between them than exists between a pair of brothers or sisters. An elder sister is closer to a young brother than an elder brother is to his younger sister. Because they learn many of their roles from their mothers, girls spend more time with them than do boys. An elder sister, therefore, has a greater share in the upbringing of a younger brother. The young boy is likely to still be in the household when his elder sister marries. At this stage her identity as a married woman reinforces the social distance of the avoidance/respect category and the young brother has no place with his sister and her husband. A junior sister, by contrast, is a joking partner of both the young spouses and spends a good deal of time with them, reinforcing the bond of same-sex siblinghood. Being the classificatory wife of her brother-in-law facilitates emotional identification with the elder sister; her tender age obviates possible strains of jealousy and the relationship between sisters is very close indeed. Brothers develop emotional interdependence in the play group and in the period of winter isolation when the elder plays his part in the younger boy's long process of learning to be a hunter. Their relationship is further stabilized by the prospect of possibly living out their adult lives in the same band, whereas sisters are more likely to be parted and to set up their respective neolocal households in different bands.

Although there is a sense of unity within the sibling group, which distinguishes its members from other kinship networks, the closest relationships are between same-sex siblings.

Cliques

In the band encampment huts or shelters are arranged in distinct coteries of from two to seven huts each. I have termed such a cluster of households a clique (there is no G/wi term). Cliques are unstable groups, which may undergo partial or complete reconstitution with each move the band makes to a new campsite. However, some may remain constant for several successive shifts. There is no apparent structural determinant of membership. Cliques consist of a seemingly random range of kin and friends of all ages. The only pattern that I could discern in the membership changes of cliques was that certain households never appeared in the same coterie together; they had been members of the same cliques but at different times. There was no structural correlate in this negative pattern. The only criterion of membership that I could discover was that, for the duration of the clique, the members had a preference for one another's company.

166

Within the clique there is a higher rate of interaction than there is between coteries. Cooperative tasks involve clique colleagues more frequently than outsiders, but the latter are commonly included. Property revolves more rapidly within the group, as the frequency of lending and giving is higher than it is between cliques. The closeness of the huts facilitates communication to the extent that eavesdropping is practically unavoidable. Conversation, gossip, and opinions are constantly exchanged in the late afternoon and at night when everybody is back in camp. The clique, although temporary, is a cohesive group within which there is a high rate of communication, shared preferences of company, and a common interest in cooperative tasks. The high rate of circulation of goods and services creates and reflects a network of reciprocal obligations. Opinion rapidly crystallizes in each clique, and in political processes, when band opinion is polarized, the cliques emerge as the poles. As consensus is approached, cliques function as subunits of agreement, within which the diverse strands of argument are ordered and simplified, clarifying the issues before the band.

The dissolution or reconstitution of a clique is not customarily preceded nor accompanied by any apparent friction. Composition is decided before the band makes its next move, agreement being reached in a seemingly casual and relaxed fashion by the simple announcement by a husband or wife that their household will be "next to" another.

In years of exceptional drought when the depleted territorial resources will not support concentration of the band into a single camp, the households coalesce into a number of cliques, and these live as tiny bands in one territory for the duration of the drought, each clique geographically separated from the others but maintaining contact through intermittent visits.

Band politics and leadership

G/wi bands are both open and egalitarian communities. The dispersal of households into isolation during the winter and succeeding months of almost every year imposes on each the necessity for self-regulation and bestows a great measure of autonomy on each household during that time. The problem in band politics is to combine that autonomy with band solidarity and retain cohesion in the polity. This is an aspect of the larger problem: The G/wi must steer a mid-course between making band life attractive enough to draw the members together but not so attractive as to prevent their drawing

The first joint camp after separation. The trees are still leafless, providing little shade. For this reason, the shelters are more substantial than is usual so early in the season.

apart again. A centralized, hierarchical structure with specialized personnel and roles would be unable to function when the band separates. As institutions tend to generate subsidiary needs, which serve to justify and perpetuate the institution, the households of the band would presumably develop some dependence on a centralized political system. Dispersal is an ecological necessity in the seasons of scarcity if bands are to maximize their size during seasons of relative plenty. The process would be inhibited by dependence on a centralized political system.

In the absence of specialized structures, there is no formal feature of G/wi social organization that might lend distinctiveness to political processes and enable one to demarcate their field of operation. Rather, I have relied on Morton Fried's definition, "Political organization comprises those portions of social organization that specifically relate to the individuals or groups that manage the affairs of public policy or seek to control the appointment or action of those individuals or groups" (1967:20–21). It follows that political processes are those by which public affairs are managed and those by which the managerial personnel are appointed and controlled. This definition is more helpful than many others, but as I indicate below, its application to G/wi bands is not without problems.

Such matters as the migratory program, the timing of separation into households, and the place of re-forming the dispersed band are clearly all matters of public policy. A proposed marriage could be seen as the proper concern of only the two households involved. Yet the opinions of other band members influence the parents in reaching their decisions, and the marriage would not be agreed to if band members were opposed to it. In a community as small and intimate as a band, the parameter of affairs of public policy intrudes far into what a larger-scale society would regard as the domain of private decision. That which is not public is permitted to be private by a public conspiracy not to proclaim cognizance, for there can be little of any individual's doing that escapes the vigilance, close concern, and profound insight of his band fellows. The comparatively rudimentary technology of the G/wi and the pressures exerted by the environment do not allow as wide a range of alternative responses as would be available in a more generous environment or to a people having a more powerful technology.

The defined political field is therefore atypical in that it includes much that is not usually considered to be of a political nature in more complex, clearly differentiated social systems and also, because of lack of effective choice of action in many situations, excludes much of what is normally considered basic political activity. Location of the boundaries of the field of politics is frustrated by the fact that under one set of circumstances a matter may involve only two or three persons in its resolution and decision. Under different circumstances, an apparently similar matter involves every household in the band.

Decisions affecting the band as a whole are arrived at through discussion in which all adult and near-adult members may participate. Discussion seldom takes the form of a single, set-piece debate. Much of the groundwork before decision is covered in the course of ordinary conversation between friends, hunting partners, and clique neighbors. If only one clear-cut course of action merits serious consideration, the casual exchanges routinely lead to a decision and perhaps function more as announcements of concurrence than processes by which agreement is reached (i.e., agreement is automatic but nevertheless requires some form of declaration in order for it to become the basis for action). Where the matter is more contentious or confused and factions emerge, protagonists will involve others by airing their views before a wider audience. The behavior of the onlookers gives a more or less clear indication of the inclination and strength of sentiment in the band as a whole. This is both "testing the wind" and influencing opinion. There are many ways of doing it: a quiet, serious discussion with one or two key individuals within the hearing of a few band fellows or a long campaign of persuasion in which the case is put together, piece by piece, allowing time for each to settle before placing the next. Or else a public, but ostensibly private, harangue is contrived by loudly addressing a friend and making sure that the whole camp can hear, that is, talking at rather than talking to. This ploy of the "forced eavesdrop" avoids direct confrontation with the opposition who would be guilty of bad manners if they were to join in the "private" conversation. However, opponents are free to resort to the same device. The band may then be treated to the occasionally comic spectacle of two sets of orators putting forth their conflicting arguments, each pointedly ignoring the other but striving desperately to avoid breaches of both logic and etiquette in their attempt to coherently answer point with counterpoint without being seen to attack directly.

The spectrum of audience response to all these preliminaries is equally broad. Some express their feelings tacitly, signaling assent, opposition, or indifference to the speaker's argument by facial expression, bodily attitude, or gesture. Others are more explicit and answer with murmurs and grunts or echo the last phrase of the utterance to show their support.

The time taken for discussion is naturally limited by the urgency of the matter under consideration, that is, the need to arrive at conclusive agreement before the passage of time and events closes off an option. Such limitations are clearly recognized by the band. Less urgent matters can be debated over a longer period of time and discussion is then intermittent, with the subject cropping up from time to time until a satisfying solution is reached.

Leadership in the band is apparent at all phases of decision making. Leadership may be measured as the extent to which an individual's suggestion or opinion attracts public support and is thus exercised in the initial stage (in which somebody identifies and communicates the existence of a problem that calls for a decision) as well as during the subsequent steps toward a final decision. In the main, leadership is authoritative rather than authoritarian; knowledge and experience of the matter under discussion and firmness of personality are characteristics that win most support. Although in themselves these are prestigious qualities and success in promoting a particular argument confers further prestige, such prestige is never sufficient to occasion an "overflow" into habitual success. Expertise in one field of activity may be seen as not at all relevant to another field, and even in matters that are quite closely related, leadership shifts unpredictably among acknowledged experts with the occasional inclusion of a "dark horse." The emotionally calm atmosphere of many discussions and the general lack of competitiveness partly explain the readiness to separate idea from identity. It often happens that the suggestion finally adopted is the one initially voiced by someone who took no further part in the proceedings, leaving it up to others to take up and "push" his or her proposal.

This is not to say that passion is unknown: Contentious matters do stir speakers to emotional oratory, and a single dissenter can shift an apparently decided band to another way of thinking. But the band is reluctant to come to decision under the sway of strong feelings; if discussion becomes too angry or excited, debate is temporarily adjourned by withdrawal of the attention of the calmer participants until things cool down. Withdrawal is not usually physical – to get up and move away is too explicit a gesture of rejection. It is, rather, an auditory withdrawal. Members signal their lack of sympathy with the heated mood by affecting preoccupation with other matters. It must indeed be frustrating to find one's fine flow of rhetoric washing unheeded round a woman busying herself with an apparently well-ordered cooking fire or wasted on a man suddenly absorbed in microscopic examination of an invisible thorn embedded in the sole of his foot, but such inattentiveness is not overtly rude. One cannot castigate this sort of absentmindedness. One must simply "chew the teeth inwardly" or try a more winsome appeal to straying minds. Auditory withdrawal is also made from a speaker who persistently pursues an unacceptable argument, leaving the bore high and dry with neither support nor legitimate cause for complaint.

Social control

As I have indicated, the processes of social control merge with political processes when control becomes a public matter. This may happen when the whole band is directly and immediately affected in the affair, or when it has concerned itself with, or intervened in, a situation stemming from the action of one or more individuals in which the band sees itself potentially concerned in the consequences of the individual's action. The political complications of openness of the community in relation to decision reaching and enforcement have much the same consequences in the field of social control. Forceful coercion is impracticable and, furthermore, it would be difficult, if not impossible, to obtain restitution for a wrong committed by an absconding member. His departure would weaken the band and possibly prove to be of greater harm to the community than was the original offense. Grievances must therefore be redressed at an early stage of development before they become serious enough to cause a lasting breach and before excessive damage is occasioned by the wrong. Hence grievances need to be formulated in terms of clear and accepted principles and aired when the level of conflict is low and its extent small. An aggrieved member should be free of inhibition in his expression of discontent and should be able to voice it in a way that can be easily and unequivocally understood.

The kinship system fills both these needs. In its normative aspects, it provides criteria by which actions can be judged as right or wrong. This is not to say that the kinship system furnishes an *en tout cas* cover of legislation for every eventuality; the values and ethics of the G/wi, as do those of every society, limit the field of legitimate grievance and, hence, the accepted causes of conflict. (To clarify this point by illustration: However much my conduct may offend the sensibilities of my rural neighbor, he has no legitimate grievance – and no accepted remedy – until I do something that is a recognized infringement of his rights.) In the G/wi case, the grievances regarded as legitimate are related to the rights and obligations governed by the kinship system or for which it provides a model.

The behavior appropriate to the joking relationship permits free and trenchant public criticism of the actions of a joking partner and imposes an obligation to accept the criticism without the kind of resentment that might exacerbate the conflict. If a joking partner's action gives offense and this can be framed even in the most generalized context of kinship rights, obligations, and expectations, he can be taken to task. If he has a defense, it can be raised and the matter argued out in public.

As the nominal hunter of the game animals that I shot, ≠xwa: was responsible for dividing the meat. One evening, after he had given out portions of a gemsbok I had killed to compensate a group of men for the time they had given me, N/udukhwe suddenly began talking very loudly from his corner of the camp about the parsimony of "some hunters." He then addressed himself to a man seated near ≠xwa: and me. "There are men so mean as to deny meat to a *gjiba:xuma* [younger brother] who has helped all day with work." Other conversations died away and the whole band stopped to listen. "There is a man who stints when he should give." The man near us replied that it was shameful that such stinginess should occur between brothers (≠xwa: and N/udukhwe were classificatory brothers). Launching his own tirade, ≠xwa: directed his words to a man sitting near N/udukhwe: "People should not be hasty; it can happen that a man has much meat to divide and many portions to give out. Perhaps he pauses, seeing that other men want to rest awhile." Gjiudwe retorted, "Perhaps a man who is tired is also hungry and it is this that makes him hasty." Charge, defense, and countercharge

were elaborated; $\neq$xwa: was not very quick-witted and came off sec-
ond-best in nearly every exchange. It was obvious that he had forgotten to give meat to N/udukhwe, but he was too vain to admit his
mistake and not nimble enough to think up a face-saving excuse that
carried any credence. In the end he called out to another man,
Djedo'o (his closest friend), that he had nearly finished distributing
meat and wanted help with carrying the last portion. (This was
pretty lame. He had, in fact, given out the last piece some time previously and hadn't needed any help. However, the jeers that I expected did not come, perhaps because nobody wanted to detonate
one of $\neq$xwa:'s outbursts and fits of sulks.) Djedo'o came over and
$\neq$xwa: quietly asked him to give up some of his portion. This was
added to what $\neq$xwa: had kept for himself and the two men took the
meat to N/udukhwe. In neglecting N/udukhwe in the first place,
$\neq$xwa: had simply been forgetful and I am certain that there was no
malice in the oversight. His faux pas had cost him his share (and
Djedo'o some of his), but nobody seemed at all sorry for him. He had
made a silly mistake, compounded it by not admitting his error, and
had paid for it.

Verbal duels are generally more direct (N/udukhwe had shown re-
straint toward his classificatory elder brother by addressing his complaint to a third party and avoiding direct confrontation) and the exchanges a good deal sharper than this. In this case the issue was
clear-cut and only really involved the two men. If a valid defense or
another point of view can be raised and argument does not lead to
resolution, the matter draws wider and wider debate until, typically,
arbitration is effected by band consensus.

The rationale of accepting a consensus judgment is not so much in
terms of whether or not the action so judged was right or justified.
Acceptance is, rather, in the terms that the band found it to have
been wrong and one either accepts their opinion or leaves for another band. It is acceptance of the community rather than of the conclusion that the community has reached. Punishment of offenses
consists of public castigation, which ends when penitence is demonstrated by apology and the making good of any reparable damage inflicted. This is not to say that a consensus judgment is always grace-
fully accepted; a good deal of grousing and grumbling sometimes
goes with it.

Refusal to heed the criticism of joking partners, lack of repentance,
repeated offenses, or ignoring band castigation can all lead to the offender's being "eased out" of the band. The offender is not expelled,
nor even ignored and cut off, but is made to feel that another band

would be preferable to the present one. The treatment is oblique and sometimes very subtle and amounts to a public conspiracy to keep the offender (or victim?) on the wrong foot and engender the type of frustration of which he cannot legitimately complain. The offender's requests for goods or services are refused, with excuses being made for refusal. His suggestions are turned down or are so misinterpreted as to make them ridiculous. Cooperative undertakings in which he participates are bungled. Much of what he says is misheard or not heard at all. I was present on three occasions when this treatment was administered. One victim was a young man with a wife and child. They were migrants of about 10 months' standing in the band, and after settling down, the man had become progressively more obnoxious to the others by loudly and obstinately voicing his opinion on everything that was happening. His joking partners sniped away at him without result, but matters came to a head over some arrows he had borrowed without returning and his refusal to make any gifts of meat to the owner, contrary to custom. The owner protested, the band backed him up with their clear disapproval, but without result, so the band decided to ease him and his family out. After a little more than a week, the family moved to another band with no apparent resentment or hostility (they spoke of making a return visit in the future) but looking forward to more congenial company.

The potential for conflict within the band is minimized by a number of factors. G/wi cosmology encourages a stoical acceptance of much misfortune as coming from a capricious and largely indifferent deity. N!adima, the creator, may act simply because he feels so inclined. He may "grow tired of your face" and visit some catastrophe on an individual or even kill him or her. Drought, the disappearance of migratory herds of antelope in the season when they are usually present, disease, and accident may be attributed to his inscrutable wish to hurt, rather than to the enmity of one individual for another.

Band territory and unexploited resources are not susceptible to individual ownership and thus occasion no disputes. Dwindling resources are not a cause of conflict, as the band either moves to a fresh campsite or splits up into household groups long before resources become so depleted that people would have to compete for them. The association of plant food with marriage restricts conflict over this commodity to the confines of the household. The obligation to share the majority of other possessions virtually defines theft out of existence and restricts conflict over these possessions to the kinship frame of reference.

Abhorrence of any violence that goes further than the domestic exchange of blows effectively prevents assaults. The latitude permitted among joking partners leaves no real possibility that mutual insults could lead to any more than retaliation in the same coin. As I have mentioned, the restricted nature of interaction between avoidance relatives effectively prevents direct conflict between them.

Conflict is expressed mostly in the form of accusations of laziness, stinginess, deceitfulness, excessively bad temper, and the damaging of property. The accusations are made with implicit or explicit reference to kinship behavior, including household rights and obligations.

Men's feasts are a means of dealing with low-level conflict between two or more men. Instead of dividing portions of a large antelope among individuals, hunters sometimes hold a feast to which most or all of the men of the band are invited. The meat is cooked and each guest helps himself to enough for his and his household's needs. The etiquette of the feast demands that all attending it behave in a friendly and cheerful fashion, that is, as if they are enjoying the occasion. A good host manages to place together those who are at odds with each other and in the enforced pleasantness, the differences between the men are usually resolved. The timing of these occasions is carefully calculated to take advantage of a cooling of hostility to the point where bitterness has passed and the protagonists are really only in need of a face-saving means of mending their differences.

Exorcising dances are performed to dispel nonspecific, general tensions and to confine the extent of unresolved conflict. I described earlier how, in G/wi belief, G//awama goes up into the sky and throws down little slivers of wood or tiny arrows that contain evil. Undetectable and unsensed, these lodge in the bodies of women, infecting them with the evil. Although men are strong enough not to be directly affected, they pick up the evil from their wives, and as it spreads, the evil poisons the atmosphere of the band, destroying harmony. (I emphasize that women are not blamed for their involuntary role, nor is there any stereotype of "women's behavior" that shows the evil at work.) A dance is preceded by a good deal of discussion: "People are discontented. Tempers are short. There is sharp talk. We should dance tonight." As the dance is strenuous and goes on long into the night, sometimes even until dawn, there is frequently some reluctance, but it is only rarely that enough men and women are not persuaded into forming a group and starting the performance, which usually ends up drawing the rest of the band in. The initiative, like political leadership, circulates in the band, coming from men and women, old and young. Girls and women sit

175

around a small fire, clapping their hands to the complex rhythms of the dance and singing in shrill, yodeling cadences. The men dance around them, forming a circle. As they pass around behind them, the dancers put their hands on the shoulders of the seated women. Those who have been taught the process of exorcising and have become skilled at it absorb the evil in the women into their own bodies through their hands and arms. They say that it can be felt moving up the arms until the two strands of evil meet in the chest, where it begins to feel painful. As a man takes in evil, he begins to go into a trance. His gaze becomes fixed and unseeing, his steps falter and he has to be supported by the other men. Suddenly he is "shot" by one of his supporters. No more than a snap of the fingers, the "shot" causes the exorcisor to collapse into near or complete unconsciousness. In some men the trance is cataleptic in character, with a low pulse rate and skin temperature, whereas others have very fast pulse rates (up to 204 beats per minute) and sweat profusely. Before collapsing, the dancer may grunt and shout and throw himself into the fire or even pick up glowing coals and hug them to his chest. This elaboration is more common among the Nharo and ranch G/wi and was presumably derived from the former. (After careful examination, I could find no signs of injury to dancers who had thrown themselves into fires too hot for me to stand near without discomfort or those who had held large, glowing coals to their chests.)

After collapsing, a man is ignored by the others for a few moments and is then rather roughly and carelessly rolled away from the circle. After a few minutes he is revived. A fast, shuffling dance is performed at his feet and head to start the process of expulsion of the evil from his body. His limbs and trunk are then massaged with long strokes in the direction of his heart. The man doing the reviving frequently moistens his hands with his own sweat from under the arms and rubs it into the trunk of the "patient" to restore him. He is then picked up and stretched upward and backward, with a sudden sharp twist at the end of the lift. The reviver also blows hard into his ears. This part of the treatment must be rather painful and the shock probably hastens the process of revival.

There are two exorcising dances, the Iron Dance and Gemsbok Dance, which are similar in form (although the Iron Dance has the additional, alternative function of being the prelude to curative "surgery" in which foreign objects, associated with the patient's illness, are removed from his body). The men dance with tremendous vigor, making short stamping steps, their feet thudding hard on the sand and the rattles around their ankles and calves echoing the rhythm, and the women clap and sing as hard as they can. A performance lasts for up to ten hours in sessions of about six minutes with a short break between each. There is not only great physical exertion but also an intense degree of emotional involvement. The performers give everything that is in them, urging one another on and losing themselves in concentration on the dance and dancers. At the end they are emotionally and physically drained but also markedly more relaxed, cheerful, and confident. Dances are performed two or three times a week when conditions are favorable and food is plentiful.

Intense socialization in the virtually closed value system of the G/wi encourages conformity. The dominant values are egalitarianism, cooperation, and harmony, and relationships are so structured that interaction is maximally rewarding when these values are expressed.

I have referred to the egalitarian nature of politics and have indicated an absence of hierarchy within which members might compete for status. Competitiveness is generally discounted. There are no competitive games, the high rate of circulation of material goods in the kinship context negates the concept of exclusively owned wealth, and the substitutability of relationships outside the household militates against the idea of exclusive groupings. (I later argue, in discussing the economy, that commodities tend to move down a gradient from the "haves" to the "have-nots.") Prestige is gained from the exercise of skills and there are many opportunities for winning prestige, but the individuals are not compared and ranked by prestige. An esteemed musician stands neither above nor below a skilled herbalist, midwife, or narrator, and each comes in for his or her share of acclaim and admiration. Balance between skilled and lay persons is maintained by household autonomy. Each household commands the skills necessary to survival, and a man or woman with exceptional skill cannot bargain for power by threatening to withhold use of that skill. Recognition depends on the opportunity to demonstrate skill, and opportunities can only be provided by a cooperative lay public.

Cooperativeness is a fragile quality in an open community of autonomous households where the mechanical nature of solidarity enforces little dependence of one specific household upon another. The rewards of cooperation are increased manpower, a greater store of shared information, and larger fields of social and economic interaction. Any or all of these may become essential to survival under certain conditions, but they are, at all times, defined by G/wi values as necessary conditions of comfort and enjoyment. Seasonal separation and isolation, the unavoidable cost of a larger community in summer and autumn, are very clearly regarded as miserably burdensome. "Real life" is the united band, and the isolation phase is a relegation to mere existence. Although the G/wi can and do forgo interhousehold cooperation, there are powerful emotional and practical needs for cooperation. However, the fragility of cooperation demands that, in the long run, it be maintained in an atmosphere of harmony. There is not only the pragmatic need for harmony; harmony is a valued end in itself. Pleasure at a plentiful supply of food or a good shower of rain can only be adequately expressed by the G/wi in the context of fellowship and harmony. When good fortune befalls isolated households or small hunting groups, the reaction includes marked regret at the absence of the remainder of the band. In discussuing "good times," consideration of food and climate takes second place to the presence of ≠ēína khwena (good, pleasant people) and tswēna khwena (literally, pretty people, i.e., nice people).

The expression and validation of self-identity in terms of these values is encouraged and facilitated by the structure of relationships and by the extent to which conformity to these values assures reward. Although the value system does not totally inhibit conflict, it does keep it within safe limits, either by lessening the number of situations in which conflict is generated (and, usually, by permitting its resolution at an early stage of development) or by removing the persistent offender from the community with a minimum of disturbance and trauma.

Interband relationships

As I have explained, bands are open communities among which members may freely migrate, either permanently or (with much greater frequency) temporarily. Membership exchange on visits and as a consequence of marriage tends to occur with greater frequency between certain bands. Although I have labeled this tendency as band alliance, there is no formality and the alliance does not preclude contact with other bands. During the period of fieldwork, every band exchanged at least visitors with every other band known to me. Each band had two or more allies, which it shared with other bands to which it was or was not itself allied. The network of alliance stretches across the central Kalahari and each G/wi band is in indirect contact with all others, with other Bushman peoples, with the Ghanzi ranches, and with Bantu cattle posts and villages. Economically, the alliance network is important as the vector of imports and exports, and the territory of an allied band is a refuge in times of localized drought. Alliances also provide the principal channels of extraband interaction. Most band-exogamous marriages are between members of allied bands, and the establishment and reinforcement of kinship bonds between the bands is therefore an accelerating process that strengthens band alliances.

The majority of interband contacts are in the form of temporary visits made by individuals and households from one band to the camp of an ally and there is also a certain amount of protracted visiting, which can be considered a temporary change of membership. In some instances the visits last so long as to become a permanent change of membership. It is also common for a whole band to visit an ally en bloc. These invasions are not preceded by a formal invitation, but their appropriateness is conveyed by visitors' gossip: "The people of Y have good food this season. They long to see the people of X." The news travels back to Y: "The people of X long to see the people of Y." Visitors from X to Y are urged to bring their band back and the exchange of pleasantries is eventually climaxed by the arrival of the people of X.

The visiting band camps at the site of the host band and forms an additional series of cliques in that camp. At the interface between the two bands mixed cliques are positioned. These consist of close friends and kin from both bands. Such mass visits usually last from three to eight weeks, that is, the duration of one, two, or three successive camps. The ease with which food can be gathered is the main determinant of length of stay. If food is not plentiful, the mass visit will neither be encouraged nor undertaken, for the resources of one camp simply cannot tolerate a doubling of population unless the season is unusually good. If a long trek must be made to the next campsite, particularly if it is not in the direction of the visitors' home territory, they will not continue the visit to the next campsite but will head for home. If food is superabundant, as in a good tsama melon season, successive camps need be only a short distance apart, and it is under these conditions that visits last for as long as eight weeks.

When short mass visits are made, the politics of the joint camp are controlled mainly by the host band and the visitor status of the other band is clearly evident. The latter are consulted, but nearly all the initiative comes from the hosts. The longer the stay, the more familiar with the territory the visitors become and the clearer their grasp of the affairs of the host band grows. The distinction between host and visitor is partly obscured. The degree of openness of communities is such that the host band has, for political purposes, temporarily absorbed, but not assimilated, the visitors.

179

Visits are eagerly looked forward to and are greatly enjoyed. The abundance of company and its variety enhance the pleasure of a season of plentiful food. Nevertheless, social and subsistence resources cannot sustain such a super-band for very long without showing signs of strain. The presence of a large number of somewhat unfamiliar people brings problems of adaptation, and toward the end of a long visit, the numbers in mixed cliques decline and interaction tends to polarize according to band membership. Visits normally end well before a stage is reached at which mutual dissatisfaction or conflict would become apparent, and the alliance is not threatened by friction. There was, however, one instructive exception. In a year of exceptionally plentiful tsamas, one visiting band stayed for 3 months. Having numerical superiority and also a generally "pushing" manner, they dominated the home band after a few weeks. The remarks of joking partners across band boundaries became more and more pointed, cooperative ventures virtually ceased, and games, dances, and discussion increasingly became one-band occasions. There was even some talk among home-band members of moving away from the tsamas in order to shake off the visitors or to leave them in solitary possession. This, however, was too drastic a remedy in a year of such remarkable plenty, and the home band decided to stay among the melons and put up with the growing unpleasantness. The visit did not end until the winter frosts blighted the melons, by which time the alliance was also pretty well blighted, and in the subsequent years of my stay, contact between the two bands was diminished and diverted to other allies.

This visit and its unhappy consequences illustrate the fragility of cooperation as well as the freedom that exists for alternation between sets of relationships that is given by voluntary, open associations such as band alliances. In the normal course of events alliances are not subjected to such strains. Interaction is only initiated in response to the need for company, for exchange of commodities, news, marriage partners, or for the need for refuge in times of local drought. If an ally is unable to provide for these needs, contact is not made. If a drought encompasses the territory of several bands, one band will not strain its relationship with other bands by invading a territory already short of food, which would be a fruitless waste of time and effort. Nor will there be any need for allies to threaten the alliance by refusing help, for help will not be asked when it is known that it cannot be given.

As relationship networks, the four organized groupings of G/wi society may be compared:

1. *The household.* This is the most markedly unitary and distinctive; relationships are complementary and cooperative and only based on reciprocity to a small extent; statuses are distinctive and ranked in the parameters parent-child, sex, and relative age of sibling. The relationships undergo regular changes in the course of the developmental cycle. Propensity for alternation and substitution is minimal. Relationships are stable in their identity and personnel but not in geographical location. The relationship network is firmly bounded internally (i.e., is relatively impermeable and nonvoluntary), but there is a high individual valency for external, nonexclusive relationships. The unit has considerable capacity for independent action.
2. *The clique.* Relationships are based on reciprocity and mutual selection; cognition of the identity of the network is weakly developed and the network is usually unstable over time. Its unitary nature is only discernible in contrast with other cliques in relation to band consensus and the unit has little, if any, capacity for independent action. The network is bounded geographically but not socially, and its existence is derived from the geographical factor. For its duration, the unit has higher internal than external valency of relationships. (Note: These are characteristics of cliques in whole-band camps, not of cliques in isolation, when they approximate to small bands in their characteristics.)
3. *The band.* The network is egalitarian in nature and based on reciprocity, common interest, cooperation, and, with respect to adults, voluntary recruitment. There is a great capacity for independent action by the unit for which consensus is a prerequisite. The stability of the identity of the network is little affected by changes of personnel or by the seasonal dispersal of personnel. The network is structurally open but is of finite size and is bounded by geographical barriers to external valency. Internal valency is high.
4. *Band alliances.* The network is based on reciprocity and has a dimly perceived identity, its operations mostly taking place within the kinship framework. The network is structurally open and is readily amended or extended. It is activated sporadically in response to specific needs and generates few secondary needs. There is a limited capacity for unit action.

G/wi social organization is not formally based on the kinship system in that structurally bounded kin groups comprise the corporate groups and other social units (the exception being the household). However, the kinship system does provide the model after which social relationships are ordered. Although they are not kin groups (in the usual sense of having exclusive criteria of membership), bands are groups of kin and kindred.

181

In general terms a kinship system constitutes a cognitive map of the statuses of its participating personnel. It is a construct of a web of social bonds, containing principles of order and simplification that reduce the complexity of crisscrossing human categorizations. In ordaining relationships between individuals, a kinship system confers upon them rights and obligations that, when exercised or fulfilled, constitute paradigms of roles (see Needham, 1971:3 ff.). In addition to this normative aspect, a kinship system also has affective aspects, which, in combination, furnish frames of reference for evaluating relationships and the consequences of particular types of interaction. The relationship categories provide a ready-made basis for action common to all who participate in the system's operation. The place of each individual is known within comprehensible limits; he can anticipate and understand the behavior of others and their reactions to his own behavior. A kinship system provides both the framework and the catalyst for interaction.

The relationship ordained by the kinship system between two individuals is a corollary or a consequence of previously existing links between other individuals. Marriage and birth are events primitive to the system (see Levi-Strauss, 1969:482–490), altering prior relationships and creating new ones. These network-modifying events are constrained by the state of the network itself, that is, their probability is influenced not only by the normative aspects of the kinship system but also by the relationships already in existence in a species of Markov chain (see Buchler and Selby, 1968:58–67). Death and divorce are more or less random events in that their probability is rarely influenced by the state of the network. (Obviously, every individual's birth will be followed by his death: This, however, is a biological factor rather than a result of the operation of the kinship system.) The event of birth, on the other hand, is also the result of biological processes, but the probability of their initiation at a given time, and the relationships consequent upon the event of birth, *are* functions of the kinship system and the relationship network within which it operates.

The fictional assumption of the occurrence of primitive events may produce elaboration of the kinship system, which widens its scope to increase the personnel of the relationship network or to intensify the relations that constitute the network. Adoption, fictional siblinghood, and fictional cousinship are extensions of the kinship model based on the more or less complete assumption that individuals were involved in birth and/or marriage. Such fictitious involvement or participation has consequences or corollaries that may be wholly similar to the actual occasion of the assumed event or may bring about lesser changes in the relationship network than would the real events. The probability of assumption of fictional events as real is partly a function of the system itself and is conditioned by the structure and state of the network, that is, by the normative aspects of the kinship system and by past events in the relationship network. It can, therefore, be postulated that a given kinship system will tend to produce network states in which particular events, real or fictional, will have higher or lower probability and that particular status constellations will have greater or lesser probability, and it should be possible to calculate and interpret the demographic and ecological significance of these probabilities.

The G/wi kinship system has an Iroquois terminological system (see Buchler and Selby, 1968:221) in which there is structural equivalence of father's sister and mother's brother and formal equivalence of siblings of the same sex. Buchler and Selby (1968:225–233) have enumerated the corollaries of these features. Characteristically, the Iroquois system tends to yield a small number of distinctive categories and to have some categories that are very broad in their scope (e.g., the G/wi term *n//odima* covers 34 possible relationships and the term *g/wāma* covers 40).

These authors remark (1968:233) that Iroquois systems define only what is prohibited in terms of marriage. The G/wi vernacular model of preferential cross-cousin marriage appears at first glance to contradict this assertion. However, as there are some 12 types of relationship covered by the term that identifies the "parent's sibling of opposite sex" and 34 types in the cross-cousin instance, the vernacular model has little positive predictive value. Terminologically, the system has greater value in predicting whom an individual shall *not* marry, namely, the 45 types of relationship covered by the terms that designate "child" and "opposite-sex sibling."

The distinction between real and classificatory usage of kinship terms is obscured by the shallow depth of G/wi lineages. For instance, few older informants were able to distinguish between the real and classificatory siblings of their deceased parents. The practice by which children of hunting partners assume a sibling relationship is consistent with this blurring of genealogical detail. As a consequence of this, there is enhanced capacity for assumption of relationships that are close to the primitive relationships. The terminology perpetuates this closeness in its tendency to merge collaterals (parallel cousins) with siblings, that is, to regress to the primitive relationship. The outcome is high redundancy among the relationships covered by the terminological categories.

The terminological categories, as an aspect of the kinship system, serve partly as status markers. They indicate joking and avoidance/respect relationships and relationships that preclude marriage. The G/wi kinship system is present-oriented in that it includes neither past nor unborn generations in lineages but encompasses only living personnel. It is a means of ordering the relationships of the living in their present circumstances and tends to obliterate the details of the circumstances of past generations. Its present orientation places the system outside the relevance of "descent theory" and the lineage principle of Radcliffe-Brown (1950:Introduction).

As I have argued, there is no formal structure that meaningfully predicts the probability of marriage between specific individuals. The marriage system is an open one, which precludes marriage between certain individuals but does not prescribe which marriages shall take place. The population is dispersed in widely separated bands, the membership of which is not structurally determined. Sex ratios, age structures, and the proportions of permitted and prohibited marriage partners (this aspect is discussed in the final chapter) vary among bands. Geographical barriers make communication difficult, and the field in which a spouse might be sought is limited and demographically unpredictable. G/wi marriages carry a heavy functional load, are subject to severe strain during the privations of the isolation phase, and lack the firm social and/or economic supportive mechanisms found in other marriage systems (e.g., the Nguni institution of *lo6olo,* which functions to stabilize marriages). The G/wi marriage system meets the demographic problem by allowing a fairly wide choice of spouse (see the final chapter) but leaves the narrower selection to more discriminating, nonstructural criteria (e.g., relative age, social maturity, psychological compatibility). The system appears to achieve optimal flexibility in the order that it imposes and to favor the chances of the marriage surviving the trying circumstances that it must endure. It is, however, an unspecialized system in that it does not rely wholly on selection criteria. There are "fail-safe, backup" mechanisms in the institution of divorce, which is uncomplicated, informal, and free of stigma, and in the ease with which divorcees may remarry (note also the institutions of levirate and sororate in this respect). In the ecological situation of the G/wi, marriage is essential to survival as it prescribes complementary economic roles of spouses and imposes economic and psychosocial obligations on parents. The marriage system ensures perpetuation of the social group of husband, wife, parents, children, that is, the household, rather than seeking only to perpetuate the marriage of two individuals to each other. In other words, the system functions primarily to ensure the survival of the household, which, as I have indicated, is the most stable of the social groups.

The extrahousehold operation of the kinship system is characterized by the high measure of redundancy inherent in the large capacity of its categories. Although the system distinguishes only a small number of statuses, there is a large range of alternative personnel who may adopt each status. Overall stability of extrahousehold groups is enhanced by the ease with which individual persons can be substituted one for another. Similarly, the small number of statuses makes for simplicity and lends a markedly mechanical character to the organizational structures of extrahousehold groups. Individuals can leave or join these groups without occasioning any great rearrangement of the structures or disruption of the activities of groups. Such organizational lability greatly facilitates the frequent changes in group size and composition that occur in the execution of the G/wi strategy of population disposition.

The tendency of the kinship system to obscure genealogical detail and thus facilitate formation of kinship bonds not demonstrably founded on events primitive in the system can operate to increase the personnel capacity of the kinship categories. Potentially, the whole G/wi population could be included in any individual's kinship circle by the universalistic device of "leap-frogging" along chains of "kin-of-kin" and expressing friendships (e.g., those between pairs of hunting partners) in the kinship idiom. Relationships are then governed by the normative aspects of the kinship system. The organizational stability of extrahousehold groups is further enhanced by the increased redundancy and lability that such extensions lend to the kinship categories.

The character of G/wi bands and other extrahousehold groups unequivocally refutes that clay cornerstone of hunter-gatherer anthropology that the patrilocal band hypothesized by Service (1971:54–71) and his predecessors, from Radcliffe-Brown (1930–31) onward, is the type that will occur under stringent desert conditions.

Although Service (1971:50) has summarily dismissed Bushmen as providing unsuitable tests of his hypothesis because "most of them have long been truly subject to Negro overlords," and the remainder "are clearly refugees," the G/wi of the central Kalahari are independent and, as I have argued in Chapter 1, they are not refugees. There appears to be no valid reason why Service's hypothesis should not be tested by the G/wi case.

It is true that, at first glance, the G/wi band appears to exhibit the characteristics postulated by Service. Members express an ideal of preferential cross-cousin marriage, the levirate and sororate are institutionalized to the extent that they are verbalized as the rationale for the prohibition of parallel-cousin marriage, and the G/wi verbalize the ideal of virilocality. However, as I have explained earlier, these ideals are not realized in the manner suggested by first appearances and do not produce patrilocal bands.

Service's functional explanation of patrilocal bands (1971:34–54) is invalid for the G/wi. Hunters have less need than their foraging, gathering womenfolk for an intimate knowledge of a territory. The behavior and movements of game animals are influenced by factors that vary from day to day. Local knowledge is undoubtedly useful to the hunter, but the amount of additional general knowledge of habitat type that the hunter needs for success does not require a lifetime of local residence; it can be acquired in a few weeks. Of far greater importance is the daily intelligence gleaned from the observations of other hunters and the womenfolk.

If anything, one might argue that the extensive local knowledge to which Service refers is of more use to women, whose search is for static things, the distribution of which is influenced by such long-term factors as soil type, community composition, and water-table levels, characteristics that may well take longer to know and understand than the particular local conditions that affect game. The need for local knowledge seems to be, therefore, more an argument for uxorilocality than for patrilocality.

The second factor that Service (1971:34–54) advanced in explanation of virilocality, namely, the need for hunters to know and trust one another, applies with equal force to the needs of women, whose foraging is at least partly cooperative and who depend on one another for assistance in childbirth and in rearing their children.

The third factor, the need for solidarity in defensive and offensive operations, is irrelevant in the nonmilitaristic social world of the G/wi (and of many other hunter-gatherers, among whom intercommunity warfare is unknown).

Service (1971:60 ff.) goes on to discuss the question of territory and the patrilocal band. His point of view, expressed with some vagueness, relies on a concept of territorality that involves occupation and defense of the territory. This concept is unfortunate in that it fails to cover the instances where competition for territory and resources is regulated by nonmilitant means. It seems that Service would have been better served by Willis's (1967:102) concept of territory: "A space in which one animal or group generally dominates others which become dominant elsewhere." Willis is an ornithologist and his definition was used in relation to bird behavior, but it aptly describes the relationships between G/wi bands with regard to their territories; a visiting band or single visitor submits to the dominance of the host band by either waiting for an invitation or by seeking permission to enter and occupy the territory. Unless the word is to be given a meaning beyond that which it usually has, there is no question of defense.

Service's militantly territorial, hierarchically ordered patrilocal band would be an ecological luxury that the G/wi could ill afford. Service touches on the cost of such a band when he discusses the possible range of its size (1971:57–58). However, he does not adequately explain the likely fate of the members of a band that shrinks below the viable limit nor does he go into the question of the formation of new bands out of those that have outgrown their limits. He alludes to the demographic problem and the cost that it can incur but does not indicate the methods for dealing with the problem. I do not argue that patrilocal bands do not exist, but as an organizational strategy, patrilocality requires some means of meeting the problems of ecological and demographic variation that are likely to be encountered by hunters and gatherers in the less productive habitats.

The G/wi band is not exclusively nor even predominantly patrilocal. Rather, its composition is affected by several factors. Some may be consequential upon structural features, for example, the initial uxorilocality of a married couple, but others are informal in nature, such as the preference an individual shows for the company of a particular band's members. Membership flows from residence and not from possession of exclusive qualifications like lineage or marriage into a lineage. The band is an open community with finite size limited by the resources available to the band. It is a highly permeable community.

The politics of the band present some problems, not the least of which is whether the band *has* any politics. Max Weber's view (1966:154) was that an organization is political in character "if and insofar as the enforcement of order is carried out continually within a given territorial area by the application and threat of physical force on the part of the administrative staff." Radcliffe-Brown (1950:xxiii) partly paraphrased Weber in his famous definition of a political organization as that "which is concerned with the control and regulation of the use of physical force." Both definitions are too narrow and selective and, plainly, deny the G/wi their politics. Southall (1965:120) toned this extreme view down a bit by defining political action as that which is concerned with the use of physical force, however remote and indirect its invoking may be. The "modality of action" definitions of M. G. Smith (1960) and Morton Fried (1967) have already been referred to. They see as diagnostic that social action seeks to influence or control decisions concerning the conduct of public affairs. As I have indicated above, the G/wi notion of what is public is so catholic as to rob this type of definition of any distinctiveness and to induce in it something of the circularity present in the famous explanation of philosophy as being that which philosophers do. In David Easton's definition (1965:75) of a political system as "a set or system of interactions defined by the fact that they are more or less directly related to the authoritative allocation of values for a society" the use of the term "values" is not necessarily in the sense of "value system" but is closer to the sense in which economists use the term, that is, the price, worth, or priority of a thing or action.

Band decisions are arrived at by consensus. The fact that the band as a whole makes the decision, that is, that the adult and near-adult membership participates, does not, in itself, constitute a consensus. Consensus is not the same thing as unanimity. It is the outcome of a series of judgments made by people who have access to a common pool of information. As the word suggests, consensus entails consent to abide by the judgment made by the group. Whether or not all members participate in the process must be a matter of their free choice – the opportunity and the information must be available – but those who choose not to participate are nevertheless bound by the decision of the rest. Abstaining implies consent to judgment made in absentia.

The shared information includes the matter to be decided, the propositions relating to that matter (e.g., the proposed courses of action), and the criteria by which the merits of the propositions are to be judged (factors relevant to the problem, the objectives of the propositions, the values and the differentials in weight accorded to each of these, plus a common knowledge, or belief, of the causal and logical relationships that exist among these other items of information). Consequently, there must be not only a shared body of information but also a shared set of standards of interpretation and evaluation of that information.

Colson (1974:5) attributes to small, face-to-face communities the characteristics of dense social networks and the common expectation of the continuity of relationships over time and points out (1974:54) that these are necessary if interaction is to be of sufficient frequency and intensity to generate and communicate the requisite spectrum of information. Partridge (1971:95) mentions additional requirements: A consensus polity can only survive in a society that is stable and free of the stress of radical internal conflict and that the specifically political aspects of the society's organization and behavior must be buffered against disruptive pressures by other, widely extended areas of voluntary cooperation.

Consensus can only operate as a mode of decision making when there is general agreement about "the rules of the game" – about the way in which decisions are made, the basis for making them, and what the foreseeable extent and nature of their consequences will be for which persons.

Consensus is not reached when a majority view emerges, nor is it necessary that all should be in favor of the adopted measure. Discussion continues until there remains no residue of significant opposition to the proposition. "Significant opposition" is the dissent of one or more members to whom the proposal is not acceptable, that is, they feel themselves unable to "live with it" and are not prepared to concede the decision.

Clearly, consensus and coercion are antithetical: The element of consent in the former negates the latter and vice versa. It is also clear that the egalitarian ethos and the openness of the band are incompatible with direct coercion. There is a measure of equivalence among bands as choices of residence for the individual and his household: Membership in one is, all things considered, about as good as membership in another, and membership is a voluntary state that can be terminated by migration. Voluntary membership means that subjection to band policies is also voluntary. At the same time, the band has the potential to withdraw the normal facilities of cooperation, fellowship, and protection from an intractable member (which is to be distinguished from the actual method of dealing with delinquents, as described earlier). But it would make no sense to do this while the conventions of consensus are adhered to. My point is that threats of migration or the withholding of the benefits of band commonwealth constitute a form of coercion to uphold and abide by these conventions – coercion as a deterrent to coercion. The threatened destruction of the social fabric of the band holds more terror than the thought of physical force. In this sense, coercion is present but in nascent form as an implied consequence of the breach of the conventions of consensus.

189

The potential fluidity of personnel inhibits the development of role specialization in political processes. If, as F. G. Bailey (1969) seems to believe, exclusive political power (in the Weberian sense of imposing one's will on others despite their opposition) is avidly sought by those who participate in the processes of policymaking, permeable communities could well leave the successful chiefs with very few Indians as most would have migrated to more promising fields of competition. There may be a fallacy in the view that that kind of power is a necessary means and end in political processes, that it is always desired, and, once gained, that it further stimulates desire. Undoubtedly, there are such individuals and some political systems encourage them to be that way, but it is not a universal trait. To restrict a games concept of politics to zero-sum games in which a player scores only at the cost of his opponent needlessly excludes cooperative and symbiotic modes of political action.

Karl Deutsch (1963) defined political power as the capacity for self-determination and linked with it the concept of political creativity – the variety of means an actor is capable of devising, discovering, and employing to achieve self-determination within the confines of the system of which he is a part. In the band the exercise of political power calls for considerable creativity. Self-determination is not confined to persuading one's fellows to adopt a particular proposition. It includes finding the most acceptable pathway to the goal of consensus and winning the cooperation of others in the activities the actor wishes to undertake. It can, for instance, be politic to concede an objective (e.g., adoption of a proposal) at one level in the interest of achieving some other objective at another level of operation or at another time – gaining power in the future by forgoing it in the present.

5

Utilization of the Habitat

Band territoriality

Resource nexus

As hunters and gatherers, the G/wi do not obtain their food by manipulating the productivity of their environment. They live off its unimproved resources and are entirely dependent upon uncultivated, untended plants and wandering wild animals for their food. Their choice of location is therefore limited to those areas where plants and animals and other natural resources are sufficient for their needs.

The essential resources are:

1. An adequate variety, number, and density of food plants to provide for their needs in all seasons and over a wide range of variation of annual rainfall and other climatic factors
2. A sufficiency of grazing and browsing to attract and sustain antelope and other herbivorous prey animals
3. Trees to provide shade and to furnish firewood and timber for constructing shelters and making artifacts
4. Pans or other impervious drainages in which rainwater gathers in the wet season to provide a supply of drinking water for man and animal
5. Sufficient space to contain these thinly distributed resources in adequate quantity

To meet the long-term needs of survival these resources must be present in combination, and a lack of any one resource constitutes a limiting factor that would nullify the utility of the others. The groups of resources tend to be mutually exclusive in their distribution (e.g., good grazing is not found under good shade trees and a few esculent plants grow on the pans surrounding waterholes) and food plants are usually rather sparsely scattered. A human population exploiting these resources therefore requires a fairly large area. It is necessary, however, that these resources be in reasonably close proximity to one another if the band using them is to be spared spending excessive time and energy in searching for, collecting, and traveling between different resources.

Utilization of the habitat

The areas in which the nexus of the first four groups of resources occurs are limited in number and not all of these nexus have a sufficiently high density of resources to sustain a human population, that is, only a limited number of areas are potentially habitable. Environmental factors, therefore, exert a pressure predisposing the G/wi to concentrate themselves in the resource nexus. Hypothetically, two strategies of resource exploitation were available: Communities could migrate from one nexus to the next as resources became exhausted or each community could be stabilized about a resource nexus. The first strategy would permit the formation of fairly large communities capable of enjoying the social and economic advantages that size allows. But along with the advantages, there are disadvantages: first, the high energy cost of the frequent migrations necessitated by the rapid exhaustion of resources; second, the danger of energy-wasting competition between communities seeking to utilize the same nexus; and, third, the impossibility of maintaining up-to-date information regarding the resources available at each nexus would have made migration a risky proposition. Consequently, the G/wi follow the second strategy, that of stabilizing each community or band about a nexus over which the members have exclusive rights of exploitation. In the context of the technology they practice, the consequence of this strategy is that both the location and the maximum size of the band are nexus-dependent.

Size of territories

G/wi territoriality is not simply a matter of occupation of an area and subsequent expansion up to the boundaries of a contiguous territory. Instead, a band must fit into a resource nexus and must limit its membership to the number that the territorial resources can sustain. I selected six territories as a representative sample on the basis of their size, their resources, and their population and surveyed each within approximate limits. The data are shown in Table 8. Both mean and modal density approximate 0.07 person per square kilometer, which is probably about the norm in the central Kalahari. The territory of ≠xade was atypical in that it was overpopulated and the band was in the process of splitting in two. G!õ:sa, a new band, was still expanding into a territory that, in the opinion of its members, would suffice for the foreseeable needs of its young members when they married and brought their spouses to live with them. (Note: In this band both boys and girls intended, and were expected, to remain in their own band after marriage.) Tsxobe, lying on the edge of the mo-

Table 8. *Characteristics of the territories surveyed*

Territory	Area (km²)	Band size	Population density (persons/km²)
$\neq$xade	906	85	0.094
G!ō:sa	457	21	0.046
Easter Pan	777	50	0.064
Kxaotwe	1036	64	0.062
Tsxobe	725	70	0.097
Piper Pans	777	53	0.068

pane forest and close to the dune country, was ecologically different from the other territories and was rather richer in resources. Despite its relatively high population density, the band appeared to be smaller than the potential number that could be supported and less of the band's total territory was exploited than was the case in other bands.

Boundaries

The boundaries of a territory are roughly defined by landmarks or, more correctly, in terms of areas surrounding these landmarks. Where territories are contiguous, a "floating" no-man's land is left between them. For instance, when the G!ō:sa people are camped near their common border with $\neq$xade territory, they hunt and gather up to the border, but the $\neq$xade band avoids the area. When the latter works the border zone, the G!ō:sa people stay away from it. (Informants explained that it is foolish to crowd the country when there are other campsites available and that too many people in the same place will scare the game animals away. Mutual avoidance is, in this situation, a matter of amicable common sense.)

The exclusiveness of the rights of a band to occupy its territory and to exploit its resources is expressed in interband relationships of the sort described earlier regarding interband migration and the visits of single persons or small groups. These rights are also dramatized in girls' menarchial ceremonies, during which the girls are ritually introduced to the territory and its resources. The rights are also expressed when an infant has been away for some time and is reincorporated into the band by a welcoming ritual.

The boundaries are recognized in the statements of informants and in the behavior of people working border zones. Gathering across the border is considered to be an invasion of rights and, to my knowledge, was always avoided. Hunting across a border is also not done, although wounded game is pursued into the neighboring territory. When an animal is brought down only a short distance across the border, it is considered to be killed on home ground. Sometimes an animal that is slow to succumb to arrow poison (see later in this chapter, under "Hunting") or a wounded animal flees deep into a neighboring band's land. The hunter either gets help from that band if they are camped near the animal's path, and gives most of the meat to the helpers, or later tells the band of the incident and gives the meat to the band as a present to affirm their rights.

"Owners"

Both visitors who cross a territory en route to another destination and those who are headed for the occupying band call at the band encampment and ask permission "to stay in your country and drink your water" (even though the waterholes are usually dry at the time visits are made and will only refill some 9 months later). In each band there are individuals known as *!ū:ma* (owner, masculine) or *!ū:sa* (owner, feminine). In the vernacular account the *!ū:ma* is the original founder of a band or his male or female descendant. In practice, because descent becomes confused after three generations and cannot be sorted out, the term is applied (with appropriate gender) to any man or woman with long-standing membership in the band. The visitor or traveler seeks permissin from the "owner," who always grants it. The *!ū:ma*'s role, in essence, is that of spokesman; it does not carry any authority over band colleagues, as explained earlier in Chapter 4.

Seasonal changes

Resources

There is great seasonal variation in the amount and variety of food plants available to provide for the nutritional and fluid requirements of the band. A square kilometer that has esculent plants sufficient to feed 50 people for 20 days in May will only meet the needs of 2 persons for the same period in September. The small-community strategy of the nexus-tied band must take into account the extreme and fairly regular fluctuations of the availability of resources. The size of the residential group must not exceed a density of population greater than the subsistence density permitted by food and other resources. In other words, the population should never have less food than is needed for bare survival – a fatal predicament. But to so limit permanently the size of the residential group, that is, to a size that can be sustained with resources are at their most meager (in winter and early summer), would be to reduce it to something only slightly larger than the average household – six or, perhaps, eight at most. The minimum viable size of a group of hunter-gatherers has not been determined. Obviously, it must be big enough to contain the manpower needed for essential cooperative tasks. In addition, as is characteristic of very small groups, the long-term lack of social and psychological stimulus would generate some problems. Considering but one aspect: In a nonliterate society knowledge is stored only in the memory. The collective memory of the 20 to 30 adults in a band is much more than ten times as rich in retrievable knowledge as would be the case if there were only 2 or 3 adults in a group of 6 or 7 individuals. (Consider how the mention of a single item in a sequence triggers the recall of a whole sequence of "forgotten" information.) I can only guess that the minimum viable size of a group in this habitat would be ten people – and they would probably lead lives comparable with the misery and desperation of the Ik (See Turnbull, 1974). Their survival would be arduous and only marginally rewarding.

The residential group

The G/wi meet the problem of fluctuating levels of territorial resources by adjusting the size of the residential group so that localized population density exerts a pressure of exploitation that is below the threshold of security, that is, a density that does not exhaust the available necessary resources. In the good seasons when food is relatively plentiful, the residential group can expand to encompass the whole band. In winter the band separates into its constituent households. The effective density of population therefore changes to match the fluctuating density of food resources. The G/wi regard the joint campsite, which is occupied from midsummer until late autumn or early winter, as the normal pattern of occupation. Depending on the size of the band and the fruitfulness of the area around the site, the local food supplies last for 14 to 30 days. Food is still available after this time period, but it is thinly distributed or far away from the camp, and its exploitation is, therefore, costly in terms of time and energy. The campsite is unpleasantly polluted with household waste, ashes, and feces. The shelters are verminous. The band then selects another site (see later in this chapter, under "Economic Choice") within its territory and migrates to it, setting up a fresh camp. This process is repeated between 6 and 15 times during this joint phase.

When the drought and frosts of winter blight the food plants, reducing their amount and variety, and game herds migrate northeastward out of the central Kalahari to better pastures, the band can no longer afford to continue as a single community. The thinly distributed remaining food plants would be exhausted in three or four days, an insufficient supply to allow the band to migrate to the next campsite, which is at least 15 km distant.

It is at this stage that the band splits up into households (small households sometimes form pairs for this period). Each household goes off to an agreed-upon part of the territory and remains there in isolation until the summer growth improves the food supply to a level that will again support the higher density of a joint band camp. Dispersing and re-forming in this way is an annual cycle in all but the best, rare years of good rainfall, mild winter, and early onset of the next wet season. By this strategy the G/wi achieve the benefits of a relatively large band membership, many times greater than that which could survive as a single residential group in winter and early summer, at the cost of temporary separation and isolation.

The fact that the band is recognized as the exclusive holder of rights of occupation and exploitation of resources means that the agreed locations of isolated households are free from the threat of incursion by others from within or outside the band, and there is no threat of competition for the limited food resources of each location. (This protection is enhanced by the great difficulty of travel during the seasons of separation.)

The G/wi resort to a variation of these two patterns of occupation in years of very poor rainfall when the land is unable to support a united band for more than a week or so even in the flush season of late summer. Instead of re-forming into one community after the separation phase, the band remains fragmented in cliques (see Chapter 4), each clique occupying its own camp. This situation is represented in the midsummer camp in Figure 13. Communication between camps is maintained by a steady trickle of visitors and an interchange of members, but the hunting and gathering range of each group is planned so as to avoid overlap with the range of another.

A fourth variation of the pattern of land occupation involves the

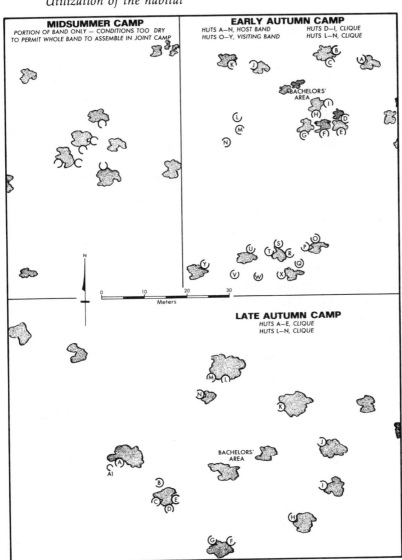

MIDSUMMER CAMP
PORTION OF BAND ONLY — CONDITIONS TOO DRY
TO PERMIT WHOLE BAND TO ASSEMBLE IN JOINT CAMP

EARLY AUTUMN CAMP
HUTS A—N, HOST BAND   HUTS D—I, CLIQUE
HUTS O—Y, VISITING BAND   HUTS L—N, CLIQUE

BACHELORS' AREA

0 10 20 30
Meters

LATE AUTUMN CAMP
HUTS A—E, CLIQUE
HUTS L—N, CLIQUE

BACHELORS' AREA

Figure 13. Midsummer, early autumn, and late autumn camps.

migration of the whole membership of one band into the territory of
another for the purpose of visiting or to seek relief from localized
drought. The politics of interband migration is described in Chapter
4; economically, the facilities of the territory are extended to the

197

guest migrants who, for purposes of resource exploitation, merge with the host band to form a single augmented population. Inter-band migration on this scale is restricted to seasons of plenty when more food is available in the gathering range of the encampment than the host band can consume in the normal three- or four-week span spent at the same site. Consequently, the presence of two or three times the normal population does not exhaust the locally avail-able food supply very much sooner than would the host band alone. At these times the move to a new campsite is initiated more by hy-giene than by hunger. Drought-relief migration, as mentioned ear-lier, is resorted to only when the host band's territory has enough food for the combined populations. Informants pointed out that to visit under adverse circumstances would be a waste of time and en-ergy, a leap from the frying pan into the fire, and would provide a pointless threat to good relations between allies. Under normal cir-cumstances a social visit lasts from two to eight weeks. In drought years the stay may be much longer; in January 1959 for instance, the G!ō:sa band remained with the Easter Pan band for the wet season and winter, only returning to their own territory when good rains fell in December of that year.

Subsistence techniques

Gathering

The collection of food plants is the most important G/wi subsistence activity. Plant foods, which constitute by far the greatest component of the diet, are eaten every day and are also the main source of fluids in all but the six to eight weeks of the year during which rainwater can be found in pools.

Table 5 (in Chapter 3) shows the average periods of availablity (ob-served between 1959 and 1966) of the edible portions of plants. There is some local and annual variation in these periods and the values given represent the periods when the foods are available in apprecia-bly large quantities. Good or early rains may extend the availability of mid- to late-summer species and early frosts may shorten the availability of autumnal species.

Esculent plants constitute the main part of the subsistence base. In terms of mass, these provide 75 to 100 percent of the diet, depending on the time of year and the fortunes of hunters. Many species are used intermittently (Tanaka, 1976, lists 67 identified and 12 unidenti-fied species) and 35 species are gathered regularly . Story (1958) reports 43 species in use. The mass of plant food gathered each day is fairly constant throughout the year at between 3.5 and 5 kg per capita. Tanaka (1976:112), however, estimated the daily intake to be 800 g. The discrepancy is perhaps due to Tanaka's having subtracted the amount of esculent plant material that is eaten to meet fluid requirements. If this is the case, our respective estimates are reasonably close.

Figure 14 represents the number of species available through the months of the year but does not reflect the amount of plant food consumed. Not all plants are of equal importance as food, but the greater number of species available in the second half of summer allows a wider choice of plant foods. A more varied diet is provided and the number of plants needed to make up the daily ration is more easily found when their variety and density are greater. Less time and effort are expended on gathering in late summer than in other seasons.

When the band is in the joint campsite phase, gathering plant foods is the work of the women of the households. They are assisted by their unmarried and dependent married daughters. In seasons of band separation, men and boys also do this work.

Women and girls forage each day for food plants within an 8 km radius of the campsite. A root, tuber, or bulb is dug out with a pointed digging stick, which is used to break up the sand before scraping it out by hand. This method is slow but efficient, for the equipment is light, easily portable, and made of materials that are

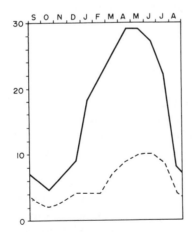

Figure 14. Monthly variation in number of species of food plants available. (Solid curve = total number of significant species; dashed curve = species of major importance.)

199

readily available and simply converted to this purpose. Furthermore, there is no call for sudden exertion; rather, there is a slow, steady expenditure of energy, which suits the small, lightly muscled women. A woman carries her gathered plant food in the lower portion of her antelope-hide cloak, which is caught up and knotted to form a bag.

The women and girls forage in groups. They cooperate by pooling information on what is to be found in particular spots, but each gathers only for her own household, and, normally, sharing only occurs between mothers and young daughters. The women leave the band encampment for the gathering grounds an hour or two after sunrise, after the morning meal has been prepared and eaten. They return some two hours before sunset in most seasons. Invalids and women with young babies or sick children come back to camp for short intervals or complete their gathering by midday and then remain in camp for the rest of the day. Except during bountiful seasons of tsama melons (*Colocynthis*), collecting proceeds slowly as the women comb the countryside for food, stopping to pick berries, then wandering a little farther and squatting to dig up a root or tuber. Sharp eyes are needed to detect the traces of dormant esculent bulbs and tubers and the barely visible signs of underground fungi. Digging requires great care, first in following the stalk down to the edible portion of the plant and then in dislodging it without damaging it, which could cause the plant to spoil. The work is not strenuous, but its lack of rhythm and the frequent changes of position, the slow, meandering searching, and the carrying of the growing weight of the day's find make it very tiring.

The amount of time required to collect a day's ration for a household varies according to season and locality. At the height of a good tsama melon season an hour's walking and picking will suffice and the time spent away from camp is short. In less favorable seasons the task may take as long as six hours. Not all of the nine hours spent away from camp is therefore taken up in foraging. Approximately half of the working day is given up to periods of rest, during which snacks of freshly gathered food are eaten. Very hot weather precludes movement. Between about 10:30 A.M. and 4 P.M. everybody rests and sleeps in whatever shade he or she can find.

The monthly variation in the number of plant species available is least during the hottest months of early summer (Table 5). At this time work is most tiring and the smaller number of available plants are harder to find. The season imposes the heaviest expenditure of time and effort in finding and gathering. For reasons explained in the

Woman gathering *Grewia* berries, which are carried home in her cloak.

following section, there is little meat to be had, fluid requirements are highest, and the supply of food plants is the most critical survival factor. The supply is usually sufficient for survival, but even with all able-bodied members of the household engaged in foraging, not enough food plants can be found to satisfy the needs of comfort. The stress of heat and sandstorm is exacerbated by the constant ache of thirst and hunger. In other seasons, except during drought years, the harvest of food plants is sufficient for the needs of the household.

Plant foods are the mainstay of the G/wi diet. In good tsama seasons, the melons and other plant foods are gathered every other day, for one day's gathering is sufficient to last the household two days. In other seasons, however, food plants must be collected each day, first because the day's take will not last longer and, second, because plant foods other than melons quickly spoil after gathering for there is no way to preserve them. They are consumed at the evening and morning meals following collection. The hunters' contribution to the diet is irregular and periods as long as 2 months may pass without meat; at these times the household relies completely on plant foods. As the principal food and main source of fluid, the supply of plant food is the chief factor in determining the pattern of routine migration within the band's territory and the selection of household campsites during the phase of separation.

Of the 30-odd plant species regularly used by the G/wi, 13 are of major importance and there is a scale of preference among these. Three of these preferred species, namely, the sweet, sappy berries of witgat, or shepherd's bush (*Boscia albitrunca*), the truffle (*Terfezia*), and the wild plum (*Ximenia caffra*), are normally abundant for only short seasons. Two other important species, *Coccina rehmannii* and *Raphionacme burkei*, which store fluid in their large tubers, are available all year round, but these are conserved and heavily exploited only during early summer when they become the main source of food and fluid. Morama beans (*Bauhinia esculenta*) have a limited distribution, being found only north of Deception Valley; they are available to southern bands only when climatic conditions permit the long journey to the morama country. There is, therefore, a real choice only among the seven remaining species. An abundance of tsama melons is preferred to any other plant food, but a plentiful supply of *Vigna* roots is preferred to a small or widely dispersed growth of tsamas. Raisin berries (*Grewia* spp.) are ranked above gemsbok-cucumbers *Colocynthis naudinianus*), which, in turn, are preferred to bluebells bulbs (*Scilla* sp.). Bride's bush beans (*Bauhinia macrantha*) are available for a long season, but the G/wi prefer to exploit the crop briefly and intensively, moving into an area in which they are plentiful and stripping the bushes in a short time, usually between the end of the raisin berry season and the best part of the tsama season. Bride's bush beans thus have a brief ascendency over *Vigna* roots and bluebells bulbs. However, even when a particular species has a good season, the G/wi do not restrict themselves to a diet of that species only but supplement the seasonal staple with whatever other plant foods are to be found in the area surrounding the camp.

The ranking of plant species in order of preference is expressed in the tactics of migration and daily collection, that is, the direction of movement is toward the highest-ranking available species. The criteria according to which species are ranked are, in order of importance, their thirst- and hunger-allaying properties, the ease with which they are exploited, and their flavor. Ease of exploitation is a function of the energy cost of obtaining a day's ration and varies with the location of the edible portion of the plant (aboveground, within or beyond reach; the difficulty of detaching the edible portion from the plant; if below the surface, the depth at which the edible portion lies). Other criteria include the size and weight of each portion, the density of distribution of the edible material on the plant, the density of the distribution of the plants themselves, and the ease or difficulty with which the edible portion is prepared for eating. The G/wi are wasteful of space and energy in that they do not strip unwanted material from bulbs, tubers, and roots before carrying them back to camp. To do this would, in my estimation, reduce an average load by about 30 percent. No reason was given and none was apparent to explain this practice.

Delicacies exist among plant foods and are enjoyed and valued but not to the extent of sacrificing "bread-and-butter" carrying space. No intoxicants or narcotics are used. Dagga, or marihuana (*Cannabis sativa*), is known and is said to have been used in the past, but it was not used in my experience.

The amount of plant food gathered in a foray is no more than is required to feed the household, any bachelor who might be a guest, and any aged dependents. As I have indicated, the amount consumed in the camp averages between 3.5 and 5 kg per person. A day's take may weigh, therefore, as much as 28 kg (i.e., before waste material is stripped off) to which must be added the weight of a good load of firewood. After a day's gathering a woman ends up with a load that is not far short of her own weight. In addition to the food consumed in camp, there is an unknown and variable amount eaten as snacks by women when foraging and by men when hunting.

Hunting

The image that popular literature has created of Bushmen as inveterate eaters of prodigious amounts of meat is false as far as the G/wi are concerned, and Lee (1969) firmly established that it is also not true of the !Kung.

Meat is certainly highly prized as food, but hunting is hard work and is by no means always successful. It is neither a full-time nor even a consistently followed occupation among G/wi men but is subject to seasonal restrictions, competing demands on available time, and occasional reluctance to take opportunities that present themselves.

The table of kills (Table 9) is the calculated average of the various species bagged in a year by the 80-member ≠xade band, which included 16 active hunters. I could not observe each kill because of my periodic return to home base to replenish supplies and because, of course, I could observe only one household at a time during the phase of band separation and household isolation. However, the periods missed in one year were covered in other years and I was able to fill in the gaps with the hunters' own reports, which were reasonably accurate.

I have some misgivings about the meat values that I have assigned to the various species, which are largely derived from von La Chevallerie's lists of carcass weights (1970:73–87). He calculates average dressing percentage as being from 55 to 61 (1970:80) to which I have added 7 to 8 percent – the edible portions of head, feet, knees, entrails, blood, pluck, and marrow (derived from Beattie, 1971:432, and referring to cattle but close enough to the values for large antelope). As the size and condition of individual specimens vary greatly within a given species, the figures are only approximations of true meat values. It was beyond my means to weigh the total edible portions of a kill – larger cuts from a carcass are fairly easy to determine, but I could only guess at the amount of meat contained in a roasted gemsbok head.

On this rather haphazard basis I conclude that the overall average annual consumption of meat is 93.14 kg per capita. (Adult consumption is slightly higher.) Tanaka, for the same band, recorded a much smaller number of kills and calculated meat consumption at a rate equivalent to 81.92 kg per capita annually. His observations covered the period September 1967 to March 1968 (Tanaka, 1976:111) when the central Kalahari had been subjected to nearly a decade of drought (Brown, 1974). Although my fieldwork also covered a long period of drought, it commenced toward the end of a decade of above-normal

Table 9. *Estimated modal monthly totals of animals killed by ≠xade band*

Species	Meat yield (kg)	Sept.	Oct.	Nov.	Dec.	Jan.	Feb.	Mar.	Apr.	May	June	July	Aug.
Giraffe	400	—	—	—	—	1	—	—	—	—	—	—	—
Eland	267	—	—	—	1	2	1	—	1	3	1	—	—
Kudu	111	—	—	—	—	—	1	—	1	1	—	—	—
Gemsbok	77	—	—	—	3	2	4	2	1	—	3	3	2
Hartebeest	45	—	—	—	1	—	—	2	1	2	2	2	1
Wildebeest	86	—	—	—	1	1	—	3	2	—	2	2	1
Springbok	13	—	—	—	1	4	2	6	6	4	6	2	1
Duiker	6	12	10	12	4	5	4	3	7	2	3	1	2
Steenbok	4.5	8	10	15	3	3	4	3	6	8	2	4	2
Springhare	0.7	3	—	8	32	30	24	28	30	26	22	15	4
Porcupine	4.5	—	—	—	—	—	—	1	1	1	—	—	—
Warthog	28	—	1	—	—	—	—	—	1	—	—	—	—
Fox	2	2	1	4	—	—	—	2	4	—	3	1	3
Jackal	2	4	—	—	—	—	—	—	—	1	2	2	3
Rodents	0.06	15	15	30	30	20	30	30	30	40	20	20	15
Birds	0.22	16	16	20	8	18	22	13	14	6	12	4	12
Tortoises	0.11	—	—	50	90	90	90	80	40	—	—	—	—
Snakes	0.2	—	—	6	4	8	4	—	8	10	8	2	—
Bullfrogs	0.1	—	—	1	2	15	24	25	—	—	—	—	—
Invertebrates	—	—	—	1 kg	5 kg	169 kg	107 kg	1 kg	—	—	—	—	—

rainfall, and I have reported elsewhere (Silberbauer, 1965:22) the drastic reduction of herbivore populations that occurred in the mid-sixties. I believe that Tanaka and I were looking at markedly different situations and that his much lower values are consistent with the decline in game numbers after the long drought.

The pattern of meat consumption is affected by seasonal fluctuations and by the day-to-day irregularities in the supply of meat. The G/wi eat large but not gargantuan quantities of meat at irregular intervals, finishing off a carcass in a day or two in hot weather. When the weather is cooler and meat does not spoil so rapidly, a large antelope may last as long as six days (by which time the last cuts are distinctly noisome).

The hunter's techniques include, in order of frequency of use, shooting with bow and poisoned arrow, snaring, catching springhares by means of barbed probes thrust into warrens, running down, spearing, clubbing, and meat robbing.

The G/wi bow is a light, tapered single-curve stave made from the wood of the raisin bush (*Grewia flava*) with a bowstring of two-ply spun sinew taken from the back muscles (*longissimus dorsi* and *costarum*) of eland for preference but of gemsbok if eland cannot be had. The stave is bound at both ends and in the middle with flat sinews (from the *splenius*) of gemsbok. A short stick in the form of a "thumb" is let into the foot of the bow and is bound fast. An eye is spliced into one end of the bowstring and is looped over the head of the stave to anchor that end of the string. The free end is passed between the "thumb" and bowstave and wound around the foot of the bow. Twisting the windings tightens the string, and pushing them against the "thumb" pinches the string and stops it from slipping. Correct tension on the string is recognized by its musical pitch when plucked. A tuned bow has a weight (pull) of about 9 kg at its draw of some 20 cm, which will carry an arrow beyond the 100-meter mark, but accurate range is limited to about 25 meters by the characteristics of the arrow.

The arrow has four parts: head, sleeve, link shaft, and main shaft. Heads may be piles (simple point) or broadheads (barbed). No. 8 gauge steel fencing wire of 4.1 mm diameter is the usual material for broadheads these days, but the ancient art of making them from bone has not been lost. A 12 to 13 cm length of wire is heated in a pit of wood coals raised to their maximum temperature by means of a blowpipe until the wire takes a bright red heat. It is beaten into shape, using one block of quartzite as an anvil and another piece as the hammer. One effect of my presence was that heavy pieces of steel, such as abandoned kingpins from my lorries, replaced the stone tools more and more commonly. Some men have acquired files for finishing their arrowheads, but most can give good points and edges to their arrows with the crude stone tools. Fencing wire is a more durable material than bone and, because of its malleability, takes a sharper edge. It is therefore the preferred material and is a valued exchange commodity.

The sleeve of the arrow is a short tabular section of reed, which fits over the shaft of the arrowhead and over the point of the link shaft, connecting the two. To prevent it from bursting under the force of impact, the sleeve is bound with sinew and served with adhesive made from the gum of yellowwood (*Terminalia sericea*).

The 5 cm spindle of hard wood or bone forms the link shaft. Its rear point fits snugly into the bound main shaft but not so tightly as to prevent it from being detached, for the function of the link shaft is to allow the main shaft to come away from the arrowhead once it has struck home. The precious main shaft can then be recovered intact, and the wounded animal is also prevented from using this shaft's leverage to remove the arrow and its load of poison.

The main shaft is about 45 cm long and is made of reed or the thick culm of one of the suitable grasses. As none of these is common in the central Kalahari, wands of *Grewia flava* are sometimes used as substitutes for the preferred main-shaft material. The arrow is not fletched or flighted in any way. Being light, it has small momentum and impact energy and is necessarily a short-range weapon; I doubt whether the increased accuracy that flights would give would appreciably increase the efficiency of the arrow.

The purpose of the arrow is not to inflict a mortal wound but to inject into the target a load of poison that, in time, will kill it. Poison (*//uadzi*) is made from the larval grubs of the chrysomelid beetle, *Diamphidia simplex*. The cocoons are found buried 20 to 25 cm beneath corkwood bushes (*Commiphora* sp., probably *pyracanthoides*, but cf. Story, 1958:28–29) and are left intact until needed to arm an arrow. Then the yellow-pink larva is carefully extracted and crushed, but not broken, by rolling it between the forefinger and the palm and by tapping it with the thumbnail. When the entrails are reduced to a pasty consistency, the head is nipped off and the insides are squeezed out and dabbed onto the shaft of the arrowhead behind the barbs, (which are left without poison to minimize the risk of accident). It usually takes eight larvae to arm an arrow, the successive

207

blobs building up to a smooth, spindle-shaped thickening of the shaft. The load has a dry weight of some 3 g. The spindle shape is said to stick more firmly to the shaft than would a simple cylinder. After being applied, the poison is gently warmed over glowing coals to case harden it. In the months when it is scarce (winter to midsummer), the poison is given the extra protection of a coat of *Aloe zebrina* juice to prevent damage by chipping and loss of potency through desiccation.

There is no vernacular antidote against a full dose of the poison, but informants maintained that the effects of a small dose could be reduced by chewing pieces of the bulb of the amaryllis *Ammocharis coranica* and rubbing the juice into the wound. The poison produces intense local irritation, a rapid abscessing and nausea, followed by fever and eventual coma.

Breyer-Brandwijk (1937) reports the poison to be highly toxic and fast acting. It resembles a toxalbumim, producing a general paralysis (not mentioned by G/wi informants), much local irritation, and hemolysis. It may be taken orally without ill effects. Shaw, Woolley, and Rae (1963) reviewed the literature on Bushman arrow poisons and reported their tests on 18 of 22 samples. *Diamphidia* was not tested as the sample was 42 years old. I should think it would have lost most of its potency by then. (However, a different poison that was 143 years old killed a mouse in three minutes! A caution to those who handle museum and collectors' pieces.)

To place his arrow effectively, a hunter must hit his prey in a fleshy part where the arrow may penetrate to a depth of 8 to 10 cm so that its poison may be dissolved by the animal's blood and carried throughout the body. To achieve this with a weapon of such limited range and accuracy requires considerable ability. Most game animals range too widely and shift their grazing grounds too frequently to allow G/wi hunters to predict their movements accurately enough to make it possible to organize beats in which the animals may be driven toward bowmen concealed at a prearranged spot (although this was apparently possible in better-favored parts of southern Africa; cf. Livingstone, 1857; Stow, 1905). The game must be taken where it is found, and to find it at all is a matter of luck, knowledge of the country and of the habits of the various game species, and up-to-date intelligence of their movements, which is gained from others' observations and the hunter's own interpretation of the signs of the animals' movements.

The two hunting strategies include the day sortie from camp, which is the more common, and the "biltong" hunt. (Biltong is a

South African term for sun-dried meat or "jerky.") Men making a day sortie usually hunt in pairs during the band's joint phase. Several pairs may go out in the morning, each pair having determined its direction and area of hunting in discussions of the previous night. Men, women, and children are alert to the presence of game, which is detected by sight or sound or is deduced from the state of tracks in the sand, the behavior of birds and other animals, or by other means, and all relevant information is passed on to the hunters. Such intelligence lends precision to the inferences made by the hunters in the light of their knowledge of the habits of game animals, the size and disposition of herds of various species, the state of vegetation and weather, local topography, and the season of the year.

The day's plans are formulated in the light of this general and specific knowledge and each party's intentions are fully discussed in order to avoid the possibility of mutual interference. Conflict of intention is resolved by precedence going to the suggestions of the acknowledged best hunters (see the section "Band Politics and Leadership" in Chapter 4). Such conflict is, in any case, rare, for usually no more than four pairs hunt on the same day and the hunting range covers 700 to 800 km² around the band encampment.

No informant was able to give me a connected, coherent explanation and description of hunting tactics. The following account is a synthesis of what was learned from discussion and observation of hunts in which I participated and from fireside conversation and interrogation concerning innumerable other hunts in which I did not take part. Although generalized, it is representative of G/wi hunting tactics.

A hunter carries his arrows, already poisoned, in a bark quiver sewn into his leather hunting satchel. The satchel, slung over his left shoulder, also carries his bow, club, digging stick, and a smaller quiver containing a repair kit of spare portions of arrows, lengths of prepared sinew for repairing bowstrings or bindings, a small supply of poison cocoons, and gum for fixing bindings (the G/wi soon discovered that heated pieces cut from my discarded lorry tires made a prime serving pitch and substituted this for gum whenever rubber was to be had). Loops and a small sheath attached to the hunting satchel hold a spear. A knife is carried in a sheath strapped to the hunter's waist.

A pair going out to their chosen hunting ground sets off not long after sunrise, walking at a good pace with the second man treading exactly in the footsteps of the first to minimize noise and avoid thorns. If they speak at all, it is in muted tones. Most communication

G/wi hunters.

is by hand signal. Tracks encountered on the way are commented on by means of gesture – the direction taken by the animal is shown by a wave: a fast wave for a gallop and a slow, wavering sweep of the hand for a grazing animal. The distance the animal is judged to have traveled since making the trail is indicated by the extent of the follow-through of the wave. A flick of the fingers shows that the trail is very fresh and that the animal is nearby; an upturned palm marks an old or fruitless set of tracks. There is a hand signal for each species of large mammal and for the small mammals and birds that are normally hunted – the fingers are held in positions imitating the shape of the head, with horns where appropriate, and the hand is moved in imitation of the animal's characteristic gait.

The hunters are always wary, but when approaching the area in which they expect to find their quarry they carefully and unobtrusively scout the land ahead, climbing into trees or edging over the tops of dunes to peer through the covering vegetation, always taking care not to betray their presence by sight, sound, or scent, keeping clear of the upwind side of a possible target. When the quarry is sighted, the pair makes a reconnaissance, noting the size and position of the herd (most bow-and-arrow hunting is in pursuit of large antelope, all of which are gregarious), whether it is grazing or resting quietly or nervously, the dispositions of sentry animals, and many other factors. The preferred target is an animal on the flank rather than on the downwind side of the herd – the hindmost beast is nearly always wary and, hence, difficult to approach within range of a telling shot. Herds generally move upwind and the foremost animals are likewise watchful, and, in any case, the hunters' scent is likely to be carried down to the herd before the front grazers can be approached. A lame but not emaciated animal is the best target, as it cannot run as far after being wounded as would a healthy specimen. However, a lame animal is likely to remain in the body of the herd, seeking protection in its vulnerable state. "Personality" is also taken into account when assessing the choice of targets.

Whether a bull or cow is easier to hunt depends on the species and the time of the year: Kudu bulls are more alert when in mixed herds before the cows separate to drop their calves, but cows run farther when wounded. This is also true of gemsbok. By contrast, a hartebeest cow may run a short distance after being wounded, and if she has not detected the hunters and the source of her pain, she may stop and stand for a while, offering the chance of a second shot. Wildebeest cows and bulls are equally tenacious of life, and the sex of the target is unimportant.

After reconnoitering, the hunters plan their approach, taking into consideration the steadiness or fitfulness of the wind, its speed and direction, the state of the herd, and the amount of cover lying along the vector of their interception of the herd. The factors are assessed rapidly, almost automatically, and the attack is worked out in a series of gestures and a whispered word or two. G/wi hunters can stalk at surprising speed through the sparsest cover without disturbing the

211

quarry. Bodies bent almost double, they advance in stages, taking advantage of the grass and shrubs trembling in the wind to mask their own movements. The trick is not to be noticed, which does not necessarily entail remaining invisible, as prey animals are seemingly only able to discern humans and other potential dangers by their scent and/or movement and not by their shapes, sizes, and colors. Crouching, the hunter slips from one scrap of cover to the next, until he is so close to the quarry that he has to sink to all fours. Then, when closer still, he must wriggle forward on his elbows, one hand holding the bow and the other holding the arrow, nocked ready to the bowstring. Within range, he waits to see that his partner is also in position. The commonest approach is a pincers movement, which allows both hunters the chance of a shot. Ideally, each puts an arrow into the target, thus increasing the amount of poison injected and guarding the success of the hunt against the mishap of a miss or of an arrow's working loose and falling out before the poison is absorbed. They do not attempt to shoot simultaneously; this is unnecessary, and the complexities of signaling the moment are too great a burden to add to the difficulties of stalking to within range. When both are positioned, the first hunter slowly moves to give himself a clear line of arrow flight, strains his arrow against the string, and lets fly. His partner shoots his arrow as soon after this as he can, but both hunters try to remain unnoticed by their prey, for there is the slight chance of further opportunity to shoot. Of greater importance, however, is that the herd may not take flight if no danger is discerned, and, consequently, the wounded beast will remain close by rather than career off on the start of what could become a long and grueling chase.

The hunters then note the tracks of the wounded animal for future recognition. If the day is still young, they follow on them, not too closely, for this would frighten the beast into running farther and faster. They stay well back, out of sight and hearing and let the tracks tell the tale of what is happening, how the poison is working, and whether other animals are perhaps becoming interested in their quarry. Toward the end, particularly, a predator or scavenger might be attracted to the prospect of an easy feed that a weakened antelope promises, and the lion, leopard, or hyena could make its move before the hunters make theirs. If the hunt takes place late in the afternoon, the men usually camp on the trail and take it up again at first light. If close to the band camp, however, they will spend the night at home and return with helpers the next day. They abstain from food and drink (and if in camp from sexual intercourse), as the G/wi believe

A pair of hunters following a wounded hartebeest make camp for the night.

that to partake of these is to strengthen the wounded animal, perhaps sufficiently for it to recover and escape. If the chase is long, lasting more than a whole day, they will be forced to take the consequences of eating and drinking.

When it is clear from the tracks that the animal is very ill and weakened by the poison, the hunters close in and spear or club their quarry to death. One hunter remains to guard the kill and begin the process of skinning, gutting, and dismembering the carcass; his partner returns to camp to fetch help to bring in the meat. The flesh around the arrowhead is cut away and discarded as it is invariably foul and suppurating. The carcass is first flayed and then cut through the brisket and belly. The entrails are carefully pulled out so as not to rupture any of the soft organs and sacs. Blood drains to the bottom

213

of the cavity and is also saved, either to be drunk on the spot or poured into clean sacs for later roasting at camp. The intestine is cleaned out below the duodenum, which, with the small intestine, is roasted and eaten if there are enough mouths to feed. (Utilization of the carcass is most efficient when the kill is near enough to camp to allow the hunters and their helpers to carry in not only the meat but the other portions as well.) The stomach of a large antelope is removed and placed on a bed of branches to keep it away from the sand. Even in the driest season it will hold 90 to 150 liters of fluid, which, although not very pleasant tasting, is a welcome substitute when water is scarce.

Next the fore- and hind-quarters are separated and the neck is severed behind the atlas bone. The rib cage is split down the brisket and spine and then quartered. The portions are now of a portable size. If the camp is not too far away (i.e., within about 20 km), these portions are taken home for further division. If the kill lies farther from camp, the meat is cut into strips 3–5 cm wide and hung on branches to dry for the night in summer or for a day or two in winter. It is then tied into bundles for the homeward journey.

From two to four "biltong" hunts are made between March and June. The majority of active hunters in the band make up a party and spend a week or two hunting in the remote parts of the band's territory. Operating from one or a series of temporary camps, they make daily sorties in the manner described above. The meat taken is cut into strips and thoroughly dried before being tied into bundles. These expeditions are looked on as something of a holiday for the men, who enjoy a brief spell away from their families. Nevertheless, they bring back a good deal of meat without putting any strain on the game resources within the daily hunting range. Their hunts also allow them to patrol the territory and the information they gather is of particular use in deciding the disposition of households during the coming phase of band separation.

Bow-and-arrow hunting is normally restricted to the season in which poison grubs are collected and for some six months after the collection. Poison supplies seldom last longer, and the poison itself goes stale and loses some of its strength after a time. Furthermore, it is usually too hot and trying in the early summer of most years to engage in the daylong exertion of bow-and-arrow hunting. Instead, hunters take to snaring game.

Other hunting techniques. Although most snaring is done by men hunting steenbok and duiker, adolescent boys and girls also set traps

214

for small mammals and gallinaceous birds. All use the same type of spring-loaded noose made of *Sansevieria* twine. The noose is pegged out around the bait (for small mammals and birds) or the concealed pit (for duiker and steenbok) and anchored by the trigger peg; the free end of the twine is tied to a stout stick of small sapling that is bent over to provide the spring. When the trigger peg is released by disturbance of the bait or by the weight of a hoof treading into the concealed pit, the spring flies up, tightening the noose about the neck of the animal drawn to the bait or around the pastern of the small antelope, and jerks the unfortunate animal into the air.

Snare bait is a ball of *Acacia* gum set in a place frequented by the bird or animal that is to be caught. Snares for steenbok and duiker are set across the paths of these solitary, territorial antelope or in breaks in unobtrusive barriers erected to subtly guide the duiker or steenbok onto the snare. Traplines are checked four or five times in the course of the day, for, apart from the fear of offending N!adima by causing one of his creatures unnecessary suffering, leaving prey too long in a trap may give it time to escape or provide a predator or scavenger with a free meal.

Springhares are caught by men using long, barbed probes. Several straight, peeled raisin-bush (*Grewia*) wands are scarfed together and the joints are bound, forming a supple stick 4 to 5 meters long. A steenbok horn is bound to one end as a sharp-pointed barb. The probe is pushed into an occupied warren and jiggled about until the barb hooks into a springhare. The length of probe down the hole and the turns it has taken following the passage indicate the position of the impaled occupant. A shaft is dug from the surface down to this point, and the springhare, which is immobilized by the probe and therefore unable to resort to its usually successful defense of digging deeper and deeper, is caught and killed by a blow behind the ears.

G/wi hunters occasionally club small mammals and birds to death, either striking them down with a thrown club as they pass within range or stalking up to the prey and hitting it with the club. Seldom used as primary weapons, spears are usually employed to give the coup de grâce to an animal weakened by arrow poison. Relatively slow-moving animals, such as ant bears and porcupines, are easily overtaken when encountered in open country and are either speared or clubbed. Spears and clubs are also used to kill animals that have been run down, but this technique of hunting is seldom used alone, as the chances of success are slender unless the quarry is weakened by injury, illness, or hunger and thirst. G/wi men have remarkable powers of endurance, however, and they do manage to chase larger antelope, such as eland, kudu, and hartebeest, until the animal can go no farther. A variation of the running-down technique is used when antelope stray near a band encampment; the men drive the beast into an ambush and spear it to death. Such windfalls are, however, very rare.

Meat robbing is a technique worthy of mention more for its dramatic than its economic impact. Some two or three times in a year, men encounter lions with fresh kills and drive them off by rushing up to the feeding lion and chasing it off. They then help themselves to the unspoiled portions of the carcass. The trick lies in correctly judging the moment; if approached too early in its feed, the lion will attack, and if left too long, until it is sated and lazy, it will stand and defend its kill rather than run.

Birds' nests are robbed of eggs during the laying season. Ostrich eggs are particularly valued, not only for their food content (each holds the equivalent of two dozen hen's eggs) but because they make very handy water containers when the wet season arrives some three months later. The ostrich nest usually contains a clutch of between 10 and 15 eggs of which only 2 or 3 are taken.

Large numbers of tortoises are caught during the months of summer and autumn. They are simply picked up when encountered by men, women, or children and are carried back to the band encampment to be roasted alive on the coals (which is the most effective method of dispatching these hardy creatures and is, in any case, a quick death in the great heat of the pit oven).

Termites, some species of ants, and an unidentified hairless caterpillar are the important invertebrates from a nutritional point of view. Termites can be caught only during the nuptial flight of sexuals. In the central Kalahari this swarming takes place only under conditions of high humidity, moderate temperature, and no wind, typically after a late afternoon thunderstorm. The whole swarming process is completed within half an hour, and as its occurrence cannot be predicted with precision, a band must be camped near a group of termite nests if it is to be fortunate enough to take advantage of the occasion. And an occasion it certainly is! Every member of the band old enough to walk grabs some sort of container and rushes off to the nests, yelling in the greatest excitement, to catch as many termites as possible. The insects swarm simultaneously in thousands and the air is filled with them, awkwardly fluttering up and away from the parental nest. Within a 100-meter radius the ground seethes with the grounded alates, shedding their wings, the males scurrying about in search of females, and mating couples running about, tandem fashion, to find themselves a home. There are more termites than the band could possibly eat but little time in which to catch them before the mated couples disappear underground. Shouting and laughing at themselves and each other, the people dash about, clutching at flying insects and picking up grounded ones. Not looking where they run, they crash into one another, trip over shrubs and creepers, and the containers go rolling in the sand, spilling their contents. The riot is soon over, as much an entertainment as it is a search for food.

Edible ants are taken whenever needed and are only eaten in small quantities. Caterpillars occur in sporadic and highly localized outbreaks. When an outbreak is discovered, the band usually moves camp immediately and remains at the site for up to a week. During this time there is little activity other than the daylong catching, roasting, and eating of caterpillars, which make up almost the entire diet of the band during these periods. There are only a few days between the time the caterpillars reach a size for eating and the beginning of their pupation phase, so the period of exploitation is brief. When the caterpillar season ends, the band either moves to another campsite or remains at the caterpillar site, if it is adequate in other respects, to exploit the normal resources of the second half of summer.

Utilization of prey animals

1. Flesh (/xa:sa). All the meat of prey animals is cooked and eaten except the flesh surrounding an arrow wound. Meat is roasted on coals, stewed, or chopped into mincemeat and fricasseed.
2. Skin (kho:sa). Antelope hides are made into leather for clothing and a variety of other articles. The dried, uncured hides of large antelope are used for the soles of sandals (but giraffe hide is preferred). When food is in short supply, uncured hide is roasted and ground into powder and eaten. Giraffe hide is cut into narrow strips and used as items of gift exchange. The strips find a ready market at trading stores, where they are made into stockwhips. Furry skins, notably those of jackals and foxes, are cured and made into headgear and karosses.* The hides of larger antelope, particularly those of eland, are also made into riems (kxodeg//wa), stout thongs that are eventually sold outside the central Kalahari as trek gear.

* A South Africanism for fur blankets. The word is derived from Khoikhoi *kho:* (skin, hide) > *khoros* (diminutive feminine singular), which is identical to G/wi *kho:* (skin, hide). See Pettman, 1913:253.

Liquor squeezed from rumen cud is filtered through grass into an empty tsama.

3. Head (*mha:sa*). The heads of large antelope are cut off and roasted entire in pit ovens. The brains (*!xō:sa*), tongue (*djamsa*), all the muscles, and the skin of the cheeks are eaten.

4. Rumen (*gabesa*). The contents of the rumen of large antelope is extracted and squeezed into the hand to expel the liquor, which is drunk.

5. Reticulum (*gabesa*). The reticula of large antelope are cleaned, dried, and broken up to be cooked and eaten with stews.

6. Colon (*g//ei /abesa*) and cecum (*g!usa*). These are cleaned and used as substitutes for stronger sinew bindings when the latter are scarce.

7. Lungs (*shodzi*). The lungs of antelope are treated as meat and are cooked and eaten.

8. Heart (*≠aosa*). The heart is treated as meat.

9. Blood (*!aosa*). This has been described earlier.

10. Pancreas (*g!usa*). Treated in the same way as reticula.

11. Liver (*gēisa*). This is one of the first organs to be removed from the carcass and is usually roasted on the spot and eaten by the men who are gutting the animal.

12. Kidneys (*khwētsera*). Also baked on an open fire.

13. Bladder (*/xam n!usa*). The bladders of springhares are cleaned, dried, and used as water-containers in the wet season. Bladders of small antelope are used as containers of lightweight articles. The bladders of larger antelope are thrown away as useless.
14. Udder (*pi:sa*). The udders of lactating larger antelope are regarded as delicacies when baked over an open fire. If there is milk in the udder, it is squeezed out and drunk before flaying commences.
15. Amniotic fluid and fetus (*humana*). The amniotic fluid of gravid cow antelope is drunk. If sufficiently developed, the fetus is treated as a small antelope and is eaten as veal.
16. Testicles (*ts'a:tsera*). Within the same band there are conflicting attitudes. Some boys regard the testicles of antelope (and of lion and leopard, which I had shot) as prizes and rejoice at being able to eat them; other boys are disgusted at the prospect and throw them away. Most men throw them away. I have not been able to determine whether more than personal taste and fastidiousness are involved.
17. The Skeleton. The *processus cornus* of gemsbok is made into arrowheads if no iron is available. The hyoid bone of a large antelope killed during a boy's initiation school is sometimes given to him to remind him of the school's teachings. The sixth to thirteenth ribs (*dzadzi*) of giraffes are cut up and the sternal ends are trimmed to a length of about 15 cm and used as sweat scrapers by women. The scapula (*//aoma*) of an eland is used as a chopper while dissecting the animal and is discarded when the process is complete (A. C. Campbell, pers. comm.). The humerus (*≠wama*), radius (*≠hune:sa*), metacarpus (*n/u/ama*), femur (*gjēsa*), tibia (*g//eima*), and metatarsus (*n/adekxamsa*) of antelope are split open to extract the marrow, which is either eaten or used in softening and treating leather. Marrow is also eaten after cooking, the bones being roasted in the fire before being split open.

This scale of utilization only obtains when hunting is based on the band camp and the household camp during the phase of separation; on biltong hunts many portions are thrown away because they cannot be used or carried home.

There is a marked seasonal variation in hunting activity and in the yield of hunts. Figure 15 represents modal values of meat consumed in the ≠xade band between 1959 and 1966. It was not possible to count all kills, as observation was not continuous and I was able to keep track of only one household at a time during the phase of separation. The shape of the graph could be appreciably changed by the killing of even one large antelope during early summer. In fact, two hunters of one isolated household killed a gemsbok and a hartebeest during September 1959. These animals have been excluded because

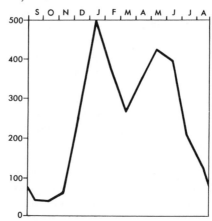

Figure 15. Estimated average daily meat consumption, in grams per person, by month.

the kills appear to have been quite atypical of this time – as far as I could discover no other households made such kills during this time and I did not subsequently see this good fortune recur. The sample is small (16 hunters) and modal values, rather than arithmetic means, are more realistic under the circumstances. The figures accord with the informants' opinions of the bags that can be expected in each month, that is, with their own concept of normality.

Water

Rainfall runoff is slight on sand surfaces. It is only after heavy and prolonged rain that waterholes receive any recharge from outside the pans and drainages. Even on the hard, calcritic surfaces of these features rainfall intensity of something like 12 mm per day is needed to cause runoff and recharge. This means that a storm must track nearly directly over a waterhole to produce the requisite rain intensity. It is, therefore, improbable that a given waterhole will receive recharge on more than a fraction of the days on which showers of sufficient intensity occur in the general area. The length of life of a pool, once it has gathered, is governed by its size, the shelter afforded by surrounding trees and shrubs, and the weather following rain days. Evaporation rates are high in the Kalahari; for comparison, Ghanzi has an annual rate of 2842 mm. Even in the humid, wet-season months it is of the order of 10 mm per day (Brown, 1974:140). To this must be added seepage loss and the water taken by animals. Taken together, this means that the G/wi have water for only six to eight weeks in most years.

Standing water, like any other territorial asset, may be freely taken by all members, provided it is used wisely and not wasted. Water is primarily used for drinking and cooking, but if a pool is large, water may also be taken for washing the body.

The availability of water is one of the factors governing the choice of campsite. As a rule, shelters are not erected within a kilometer radius of the waterhole, and the camp may be placed as far as 5 km away from it in order to avoid frightening away game animals (and risking confrontation with predators), which share the water with man. Water fetched by a member of a household is the property of that household. A person's needs determine how much of the household ration each is given; children receive larger rations because their water requirements are considered greater than those of adults. Pregnant women and nursing mothers are also accorded larger shares. Although tin cans and iron pots are used to a small extent, the main water containers are springhare bladders and ostrich eggs, which have been emptied and cleaned out. The bladders, light and compact when empty, are fragile when filled with water. Ostrich eggs, on the other hand, are bulky and awkward to carry but make somewhat stronger vessels. If the water source is fairly near, only ostrich eggs will be used, but if the waterhole is far off, the greater portability of the empty springhare bladder is an advantage. Early in the wet season when a distant pool collects some water from an isolated shower, but not an amount sufficient to justify moving the camp, a small group of young men will go off and fetch as much of this water as they can carry. The water they bring back is shared among band members, each taking a drink in turn, and is consumed on the same day. No direct reward to given to the young men for fetching the water.

A few waterholes have been improved by excavation. The most notable is the depression on the northern side of Tsho:khudu: Pan, where a tank with a capacity of nearly 5000 liters has been dug by the people of that band. In G/edon!u valley the people of ≠xade band have increased the capacity of a depression in the calcrete valley floor by excavation and by redirecting the natural drainage into the sink. Nearly all waterholes are periodically cleaned out and windblown sand and other detritus are removed. As grazing animals like to shelter under the trees that are found around virtually every waterhole, their droppings accumulate and are washed into the waterhole by the early showers. The cleaning, therefore, not only improves the capacity of the sink but also marginally improves the quality of the water. The maximum capacities of significant waterholes (i.e., those regularly used as the focus of a campsite) vary from approximately 3600 liters (G//akokum) to more than 500 megaliters. The pool life of these averages are, respectively, 6 and 24 days after filling.

Use of trees

Trees are an important territorial asset. Apart from the shade and firewood they provide, their wood is used in making several artifacts that are essential tools of survival. Wood is the only fuel available for roasting pits, cooking fires, and the smaller fires that keep household members warm in the bitter cold of winter nights. Some plant foods are indigestible and others are poisonous (notably *Coccina* roots, one of the staples of early summer) in their raw state. The absence of firewood would impose a severe handicap. Considerable quantities are burned, so the proximity of an adequate supply of firewood is a factor in determining, within more narrow limits than does the food supply, the locality of the campsite.

Its exact location and the placing of the shelters or huts within the campsite are determined by the position of suitable shade trees. Like other Bushmen, the G/wi are susceptible to sunburn and must protect themselves by spreading their cloaks or by remaining in the shade of trees. As it is awkward and sometimes impossible to work under a cloak, shade trees are essential. Shelters will not serve this purpose as they are small and not roofed in the dry seasons. In any case, guests are not received in the shelter, the sleeping quarters of the household, and much of the day is spent in group activities that preclude the use of shelters.

Although there are no evergreen species in the central Kalahari, apple leaf (*Lonchocarpus*) and shepherd's tree (*Boscia*) are seldom entirely leafless and the pattern of twigs of some of the acacias is dense enough to give a filigree shade even when all the leaves have fallen.

Ideally, each shelter or hut, or each group of shelters whose occupants constitute a clique within the band, should have a shade tree and there should also be a tree for the bachelors, who have no other shelter. The preferred site has an isolated spinney with shade for all that allows an unobstructed view of the surrounding countryside. (This often means that there is no shade left for the fieldworker's camp.) Shade trees are also handy repositories for weapons and other possessions – and safer, too, when young children are in the household.

Firewood and wood to be used in the manufacture of artifacts only

222

become individual property after they are cut, for example, one branch may be cut by one man, but other men can also cut branches from the same tree. When a tree is felled and the trunk is used to make a large object, such as a mortar or a bowl (for which *Albizia* and *Ochna* spp., respectively, are used), the waste wood may be taken by anybody else who needs it. Standing trees, thus, are not susceptible to ownership but are in the same position as other territorial resources, that is, available to any member of the band.

Trees also house swarms of honeybees (*Apis mellifera adansoni*). Although informants' statements were inconsistent and I seldom observed any utilization of this resource, it appears that nests are robbed when discovered in the normal course of felling trees for other purposes. Swarms occur most commonly in *Boscia* and *Albizia* spp., which are popular trees for artifact material. The combs are shared by those brave enough to participate in robbing these singularly vicious bees and are distributed widely among band fellows.

Rainwater collects in the hollow trunks of some witgat, or shepherd's trees (*Boscia albitrunca*). The dry, pithy stalk of the wild dagga plant (*Leonotis* sp.) is used as a drinking straw to suck out the water. Such trees are reported to other band members when discovered and are preserved. The stored water amounts to no more than a few gallons and while it lasts is drunk by all who pass by.

The fruits of such trees as *Zizyphus*, *Ximenia*, *Boscia* spp. are, like other esculent plants, harvested by all band members.

Material culture

Animal products – leather

In flaying a carcass, the first cut is made from chin to crotch, and then outward along each hind leg down the gaskin to the fetlock. The next cuts are from brisket to the fetlocks of the forelegs. The hide is peeled off, starting at the chest and working outward and backward. It is separated from subcutaneous tissue by pushing and hitting with the clenched fist and by beating the loosened portion with the butt of an axe handle, then pulling, and by cutting resistant tissue where necessary. The skin is pulled off the legs to the fetlocks and is cut above the hooves. At this stage the carcass is rolled over onto a bed of branches, which keeps it clear of sand. The skin is now peeled off from the back, working from tail to head. Incisions are made to clear the ears, eyes, nose, and mouth, and the hide comes away in one piece.

It is next pegged out, hair side down, to dry in the sun. Sharpened pegs (*/ebag//wa*) of *Grewia* wood, 25 to 35 cm long, are pushed through about 12 mm from the edge and rammed down into the sand, holding the hide smooth and taut. Straw is spread beneath the hide to elevate it from the sand, to allow ventilation, and to discourage termites and other insects, which might damage the hide. In warm, dry weather one day's exposure is sufficient to dry the hide. It is then taken up and scraped clean of blood, fat, and other unwanted tissue with a razor-sharp adze. Care must be taken with this long and delicate task; for the skin is easily pierced and spoiled. The time required depends on the size of the hide; a steenbok skin takes half a day. "If the adze bites well," a gemsbok or eland hide can be scraped clean in three or four days of intermittent work. Steenbok skins, which are used to make clothing, are also scraped on the outer surface, for the hair would be very irritating to the wearer.

After cleaning, a steenbok skin is first wetted with the juice of the bulb of *Crinum rehmannii* (*//ou*) and then repeatedly kneaded and wrung out. It is then worked between the fingers and any unwanted tissue still adhering is scraped off with the fingernails. Human urine mixed with *Crinum* juice is also used to moisten the skin. The juice of tsama melons is also used when they are in season. The skin is then allowed to dry. After a final kneading to further soften it, it may be used to make an article of clothing. In some cases it is necessary to repeat the wetting process once or twice before the skin is soft enough to use. By then it is as fine and as soft as good kid or chamois leather.

The larger hides, such as those of eland, gemsbok, kudu, or hartebeest, are first smeared with fat mixed with brains. Antelope fat, preferably eland, makes the best mixture. The oil from *Ximenia* nuts is an acceptable substitute. Along with brain matter, marrow from the long bones of antelope is added to the fatty preparation, which is allowed to soak into the hide for about three days. The hide is then put through three wetting processes, similar to those used for smaller hides. The whole curing process requires 14 to 16 days. For much of this time the hide requires no attention (e.g., during the permeating and drying stages) and the leatherworker is free to engage in other activities, such as hunting. Kneading and pegging, however, require the assistance of five or six others, as a man cannot manage a large hide single-handedly.

Most articles of G/wi clothing are made from antelope leather. Men's breechclouts (*//weme*) are made from prepared steenbok hides. The wearer makes his own garment, cutting the hide in a triangular shape to fit himself. The apex of the triangle is passed between the legs and fastened to the other two corners (which are bound around the waist) by means of a loop cut into it. The broad part of the triangle covers the pudenda and the two sides of the triangle hang down slightly between the wearer's legs. His buttocks are left uncovered. A breechclout takes about 90 minutes to cut and trim and lasts for some 12 months before wearing out. The garment is not taken off during its life but is loosened when necessary (for defecation, urination, and sexual intercourse).

The cloak (*n‖aosa*) worn by both men and women also serves as a blanket and a carrying bag. It is made from the hide of a large antelope. The skin of the upper part of the neck and the skin covering the forearms and stifles is cut off and the remainder is trimmed to approximately rectangular shape, with short projections left at the four corners. The hair may be left or wholly or partially removed – in the last case shaved with an adze to leave a decorative design. The cloak is worn over the back, with the upper projections being used as ties across the shoulders and throat. The stubs from the stifles form the waist ties when the cloak is looped up into a carrying bag. A cloak lasts some 6 months, after which it tears under the strain of heavy loads and from the abrasion of constant brushing against thornbushes. A man is responsible for providing a cloak for his wife and for replacing it when it wears out. Trimming the prepared hide takes only about an hour.

Young girls wear small fore-skirts or aprons to more or less cover the pudenda. As a girl approaches puberty, she takes to wearing a more effective covering and also dons a hind-skirt, which covers her buttocks. In addition, an apron may be worn over the fore-skirt. Both the fore- and hind-skirts are worn by all adult women.

The fore-skirt (*g!eisa*), as its G/wi name suggests, is made of steenbok skin. It is cut in the shape of a shield and short straps are sewn to its upper points as waist ties. Bead embellishments may be added. A popular type of fore-skirt for small girls, which is also worn as an outer apron by older girls and women, has a fringe made of the metatarsals of springhares, each bone attached to the waist strap by a small thong.

The hind-skirt (*!u:sa*) is made from duiker hide. It may be shield-shaped or square, measuring some 45 cm across the top (a fore-skirt is some 22 cm across the top), and is also fitted with waist ties. Some are decorated by patterns of hair left on the hide, and tassels and fringes may be sewn on.

Skirts are made by men for their wives and unmarried daughters. The time required depends on the type of skirt and on the type and extent of embellishment. Skirts become worn and begin to rot after some 18 months.

Sandals (*n//abosera*) are cut from dried but otherwise untreated giraffe hide. Eland and gemsbok hide are less durable substitutes. The sole is a single slab of hide, cut to about 1 cm greater than the outline of the foot. The hair side is worn uppermost. One strap passes through slits in the arch of the sole and up over the wearer's instep. Another passes from between the great toe and the second toe and is looped through the first strap. A third strap passing behind the ankle loops through the first strap. In putting on the sandal, the foot goes through this last, encircling strap and the toe strap fits between the toes. The sandal is fastened by tightening a knot in the first strap.

Most G/wi wear sandals only when the sand is too hot to permit the normal barefoot style (sand temperatures reach 72 °C). The footwear soon wears out, as the untreated leather is rather soft. Sandals are made by men for themselves and their families and as gifts to exchange.

Antelope leather is also used to make pouches and satchels. The hunting satchel, which has already been mentioned, is made of steenbok skin. After tanning, the hide is sewn together again. The neck is drawn tightly about the base of the hunter's quiver, and when the long ventral slit is sewn up, the hide begins to resemble its original shape. Fore and hind legs are joined together to form a strap, which goes over the hunter's shoulder. The cut between the hind legs is left open as the mouth of the satchel. A small sheath holds the blade of the spear, and loops attached to the shoulder straps carry the bow. These weapons are carried on the outside of the satchel so that they will not catch on the contents of the satchel itself if they have to be drawn in a hurry.

The thread used in sewing leatherwork is made of fine pieces of sinew, cleaned, dried, and spun together after the sewer remoistens them by drawing them through his pursed lips and twisting the strands between his fingers. An awl is used to pierce the leather, the holes being spaced according to the type of work in hand. If strength is needed, stitches are spaced about 1.5 mm apart with the precision of those done by a sewing machine. Blanket stitching is used to join pieces of leather, although the work may be oversewn if less strength is required. Faggoting is done on ornamental work where no strain will be put on the stitches. Lockstitch, running stitches, and backstitching are also used where appropriate.

In addition to the hunting satchel (*g≠ubema*), the G/wi make and use several types of carrying bags (*g/wamg//wa*), ranging in size from something more than 30 cm² down to tiny 2.5 by 5 cm sachets. They are sewn inside-out and lock- or blanket-stitched up the sides and across about half the top, leaving an elliptical opening. Straps are stitched into the edgeworking of the larger bags and twisted together for strength (which makes them cut into the shoulder when a heavy load is carried). The sides are often decorated with beadwork, tassels, and ornamental stitching. Nearly all the bags are made by men and among these highly mobile people are used by all, men, women, and children, for carrying all manner of smaller objects.

Another important animal product is beadwork (*!xamdzi*), which is made from sherds of broken ostrich eggshell water containers. The shell is broken into pieces about 1 cm² and a hole is drilled through each piece by means of an awl twirled between the palms. This takes about 30 seconds. Betweeen 120 and 150 such pieces are then strung together on a sinew thread. The assembled string is placed against a firm surface, such as the side of a mortar, and rubbed with a piece of gastropod calcrete (a soft stone). In about 20 minutes the string is ground to an even cylinder, that is, each piece is rounded off into a disc.

Strung together on sinew threads, these eggshell discs are made up into women's aprons (*g!aji*), armbands, women's coronets, decorative squares, and a variety of embellishments for clothing, pouches, and satchels. An apron measuring 22 by 28 cm and containing 4000 beads represented nearly 200 hours of work.

Beads are also used to make harnesses for babies who have not yet learned to walk properly. Stout strings of beads that pass over the chest, under and around the arms, and around the waist and each leg are connected by a string passing down the length of the spine. In addition to being an aid in teaching a child to walk, the harness provides a handy grip on an infant who wanders into danger. It is also considered an elegant decoration. A completed harness represents some 60 hours of work.

Although these beads are seldom broken, the thread on which they are strung is less durable and some beads are lost whenever a string breaks. When a beadwork article disintegrates, the beads are restrung to make some other item. As, year by year, more and more articles are made, the total store of beads mounts. They are an important medium of gift exchange within and between bands and even cross ethnic boundaries when exchanged with different Bushman and Bantu peoples, who value them as decoration. An undetermined proportion of G/wi beads come to them from the !xō, their western neighbors.

Ceramic trade beads also circulate in the central Kalahari and supplement the indigenous article. The G/wi like red, white, black, and blue trade beads equally well. Next in their order of preference come yellow and green beads. Copper beads, made by beating out short lengths of heavy armature wire and bending them to form small, hollow cylinders, are greatly liked and are worn as simple strings and as decorations on women's aprons.

Dancing rattles (/xododzi) are made from the cocoons of an unidentified moth or butterfly (described by informants as "a moth/butterfly [synonyms in G/wi] which is found on *Acacia mellifera*. The caterpillar emerges just before the wet season. It is red-brown, about 6 cm long [indicated] and is covered with hairs which sting if you touch it. In the wet season the caterpillar turns into a moth/butterfly. Many of these cover *Acacia mellifera* trees and defecate their eggs all over the tree").

The cocoons are gathered in winter. They are covered by protective hairs, which are intensely irritating to the skin. The hairs are knocked off with a twig and by rubbing with a wad of grass. The cocoon is slit down one side and the pupa is removed. Small pieces of ostrich eggshell are placed in the cocoon, which is then strung, along with others, on thongs threaded through the upper and lower points. The thongs serve to keep the incision closed, retaining the pieces of eggshell within. Some 80 cocoons are assembled to make a string about 150 cm long. Each man has two strings of rattles. They are tied around his calves for dancing, and when they rattle during dance steps they augment the rhythm of the women's clapping. The rattles are used on an average of three nights per week when the band is camped together. They are given hard use; the wild stampings of the dancing men put a great strain on the thongs, which eventually break after some ten weeks. The rattles themselves (i.e., the cocoons) rub through after a year or two of use.

Woodwork

The G/wi carve bowls (n!adzi) from the dark, fine-grained wood of *Ochna pulchra*. A section of trunk, when still green, is trimmed down to a cylinder 25 to 28 cm in diameter and some 15 cm high. The center is chiseled out and the outside pared away, leaving a curving side 6 mm thick at the rim and 12 mm thick at the rounded base. Geometric designs are burned into the outside wall. These bowls are used for eating and for storing food temporarily. Although still common, they are being superseded by enamel bowls, which are lighter and more durable. Bowl carving is the work of men, who make them for their wives and as an exchange medium. Each bowl takes one or two days. The close grain and treatment with *Ximenia* fat prevent the bowls from splitting after drying.

Mortars are made from the wood of perdepis trees (*Albizia anthelmintica*) or *Ochna pulchra*. The selected section of trunk is cut to a length of about 60 cm, barked, and trimmed smooth. It is then seasoned by lightly burning it in a fire and then burying it in a pit oven. Seasoning demands skill and experience; if the wood is not adequately seasoned, it will split during or shortly after manufacture; if the process is too rapid, the wood will burst under the pressure of steam generated within it. Because the G/wi have no instruments to measure heat, they must guess when the oven is at the right heat. Oven curing takes seven or eight hours and is usually done after the evening meal, when the oven is still hot from cooking. The wood is left buried in the sand and coals until morning to gradually cool, which is said to make the seasoning more effective.

The seasoned block is half buried in sand to hold it steady and is hollowed out from the top to within about 10 cm of the base. An axe or adze blade set lengthwise in a wooden handle, or in a gemsbok horn, is used as a chisel. The sides of the mortar are left about 18 mm thick and a flat piece of quartzite is usually set in the floor to provide a hard crushing surface and to spread the force of the blow of the pestle and lessen the probability of the base splitting under impact. Some men bind their mortars by sewing on a covering of wet eland or gemsbok skin. As it dries, the green skin shrinks tightly about the wood, lending it greater strength. A twisted thong is sewn onto this covering as a carrying handle. The manufacture of a mortar takes nearly two days, excluding seasoning time. The work is done by men for their wives and for use as exchange media.

Pestles are carved from logs of *Boscia albitrunca* with adzes and chisels; the work takes about three hours.

The G/wi use firesticks ($g \neq ag \neq abama$): The upright is a dry stick of hard wood, such as *Grewia*, about 45 cm long, which is held between the palms and twirled, with strong downward pressure, in a notch cut in a piece of dry, softer wood (*Acacia mellifera*). This lower piece (*//ane:ma*) is usually placed on a knife blade to prevent it from sinking into the sand under the pressure that is applied. Tinder (dry grass) is placed under the knife point and is ignited by the glowing grains of wood that are drilled out by the twirling upright and tipped onto the tinder. Forty-five seconds is the average time taken to produce a flame.

229

Fire is an essential tool in setting metal blades of many tools into their handles. The handle of an axe, adze, chisel, or knife (*Boscia albitrunca* is the wood usually used) is trimmed to shape, and the socket to take the appropriate blade is burned into the wood by pressing the heated metal against it.

Huts and shelters

The design of the hut or shelter (*n/usa*) varies with the season. The most substantial structure is made in the wet season. A frame of branches and sticks is made by sinking the butts 20 to 30 cm into the sand, the upper ends intertwined and supporting one another to give a rough hemispherical shape to the whole. The frame is thatched with clumps of grass pulled up by the roots. Unlike most thatchers, who place the tips downward, the G/wi point the roots downward. The weight of the clump above therefore anchors the tips of the one below, packing the thatch more securely. In most households the greater part of the work is done by the wife, with help from her friends. However, there is no rigid division of labor and in other households the husband may do some or all of the work.

With the passing of the wet season, the shelters become less substantial and are intended to shelter the occupants only against wind and sun. When there is no probability of rain, the shelters are not roofed but are simply semicircular enclosures made of leafed branches stuck into the sand. Gaps are plugged with clumps of grass. After winter, when the wind becomes boisterous, the floor of the shelter is excavated to a depth of about 30 cm to give better shelter against the stinging blasts of sand.

Shelters and their forecourts are swept several times a day. A ridge of detritus soon forms at the perimeter of the forecourt and this remains to mark the site long after the thatch and wooden framework have decayed to dust. The entrances of shelters and huts are faced away from the prevailing wind; that is, they face southwest except for the early part of the wet season when they are faced away from the expected westerlies. Some households face their entrances north in winter to avoid the occasional bitter south wind, which has the advantage of catching the northern sun.

Metalwork

Soft-iron rod is an important material in the manufacture of many G/wi implements and weapons. In the Ghanzi ranching area rods of 9.5 and 16 mm thickness are commonly used to make branding irons, and stores and smiths carry plentiful supplies imported from the south. Offcuts of rod circulate widely in the central Kalahari along the exchange routes. Their use in making spear blades has already been mentioned and knife blades are made in similar shape and fashion. Axe and adze heads (which are interchangeable) are forged from 7 to 15 cm lengths of rod, heated by blowpipe in a pit of wood coals into the shape of a truncated triangle. The shoulders are flared out in hyperbolic curves, widening at the flat or crescentic base, which is sharpened by hammering and then grinding in a smooth piece of stone. The narrow apex of the triangle is squared in section and fits tightly into the hole that is burned through the wooden handle. One man can forge and sharpen an axe head in about six hours. The soft iron soon loses its edge and is honed against an awl. The G/wi custom of sharply blowing across the honed blade, making a whistling sound, "blesses" the blade, they say, and keeps it sharp for a little longer.

The G/wi axe or adze (*n!abisa*), a multipurpose tool, is used in all types of woodwork, in the butchering of carcasses, the preparation of food, the working of skin, and even, in very hard ground or soft rock, as a digging tool. Every adult man has an axe, which is also used by his wife.

The G/wi awl (*g!a:ma*), which is used in leatherwork to make the stitch holes, is made from a piece of thin wire about 12 cm long. A steel bicycle spoke is the preferred material, as it is stiffer and harder than anything else available. (This unlikely commodity is stocked by stores although never a bicycle is seen.) Slivers of bone serve when no iron or steel is available. The spike is set in a cylindrical handle of wood and is sheathed in a matching, tight-fitting tube. The sheath and handle are usually decorated in continuous patterns of geometrical motifs that give the effect of being one unbroken cylinder. As already mentioned, awls are also used to hone axe and adze blades and are frequently employed to remove thorns from people's hands and feet.

Fencing wire, which is now the usual raw material for the manufacture of arrowheads (see the earlier section, "Hunting"), is also forged into keys for the *tē:kenisa* (finger piano). Pieces too short for other use are beaten out to form earrings (*n!a:na*). The design is standard among all Kalahari Bushmen; the lower end of the earring is a kite-shaped tetrahedron about 12 mm long with a slightly raised longitudinal spine. Above this is a round motif suggestive of knotted string, from which the hook or ring projects. This requires very fine work and each earring takes four or five hours to make.

Tobacco pipes are made from brass cartridge cases (Mk VII .303 fitted into Martini-Henry) and from discarded tin cans. The cans are cut and flattened and the resulting rectangular sheet is rolled into the shape of a frustum, the edges meeting in a neat folded joint. A wad of crushed leaves is placed in the narrow mouthpiece end of the frustum to prevent the tobacco from being drawn into the smoker's mouth when he sucks on the pipe.

Goods and services

Ownership

I have equated ownership with exclusiveness of use or control of use. The concept of ownership is not highly developed among the central Kalahari G/wi and I found difficulty in discussing it with them. Nor was the language very helpful; the genitive construction indicates possession as well as ownership and includes parental and other nonownership relationships. I have translated the term *!ū:ma* (and the feminine form, *!ū:sa*) as "owner," but this is misleading. The term also has the significance of "master," both in the sense of being the reciprocal of chattel and in the sense of superiority. Furthermore, the term is only applied to a limited range of all objects of ownership or mastery. There is no judicial process in which ownership is contested and in which the concept might have been formulated; in their intimate, closely shared band life ownership of any material or incorporeal thing is general knowledge and a statement of ownership is unnecessary. Because no exposition of ownership was available from informants, I have relied upon the criteria of exclusiveness of use or control of use as the key to ownership.

Undisturbed, untouched, and unclaimed territorial assets are owned by N!adima but are subject to human ownership once the process of their exploitation has been initiated. For instance, when a man signifies his intention to hunt a particular animal, it is wrong for any other man to attempt to take the animal for himself unless and until the hunt is abandoned by the first claimant.

Food. Food is the property of the household whose member collects or hunts it or who receives it as a present. Gathered plant food is normally consumed within the household of the collector, although such food is sometimes distributed to elderly and invalid people who are not household members. The plant food is taken to them and is eaten in their own huts or shelters. The only guests fed within the confines of the household are young boys and girls who are invited either as playmates of the children of the household or to encourage a relationship of boys and girls of marriageable age that is favored by the parents. Women apportion food within the household and it is their responsibilty to see that each member gets an equitable share (e.g., the mother sees that the children do not eat up their father's food). Men attend to the distribution of gifts of meat outside the household.

Only enough plant food is collected to supply the household's daily needs. Although the gatherer can confidently predict the measure of daily success, the hunter has far less control over the size of his daily bag. Because some hunters bag more than their households could possibly eat before the meat begins to spoil and others come home empty-handed, meat is an obvious exchange medium. In an attempt to maximize their surplus of meat and thus increase their quantity of exchange medium, some hunters select as their quarry the largest animal that can feasibly be hunted under the circumstances of the day. Maximization of meat surplus does not, however, entice the hunter into shooting more than one head of game. This is regarded as greediness and would anger N!adima. In any case, it is doubtful whether a hunter could successfully handle more than one head.

There are few formal arrangements governing the distribution of portions of a butchered carcass. Division of the spoils is, rather, a matter of striking an acceptable compromise between the needs of all the people to whom the hunter feels obliged to give meat.

The hunter whose arrow first struck home is the *!ū:ma* (owner) of the carcass. He takes slightly more than half of the meat and other useful portions for himself and gives the remainder to his hunting partner. If the successful arrow is borrowed, the hunter either hands over the whole carcass for distribution or makes a gift of half of his own share to the owner of the arrow. This actually only amounts to a choice between personally discharging his own obligations and making a grand gesture of magnanimity, for either way, the hunter and his partner end up with much the same amount of meat because the owner of the arrow returns to them portions about equal to what they would have kept for themselves had they not presented the carcass to him.

There are two modes of distribution: the men's feast and the presentation of individual portions. The men's feast is similar to the *anga* described by Firth (1950). The hunter alone or in concert with his hunting partner (or, sometimes, including the arrow lender) announces a feast and issues invitations. In the afternoon of the day after the kill a cooking fire is made outside the host's shelter and the invited men gather to cook pieces of meat. After they have eaten for a couple of hours, their wives and dependent children return from their food gathering and play and drift to the fringe of the gathering. Each guest is then given a portion of raw meat, which he passes on to his wife or sends home with a child. In this way the host and his associates are able to deal with a number of obligations simultaneously and also derive prestige from having provided an enjoyable social occasion. As I have mentioned, feast giving is also used as a means of restoring harmony to strained relations and the festival atmosphere of the feast is deliberately manipulated to help rebuild the bond threatened by the late quarrel.

In distributing individual gifts of meat, the size of the cuts is guided more by the needs of the recipient than by the magnitude or quality of the obligation that the gift of meat is intended to meet. The quality of obligation is a function of the relationship between giver and recipient and is expressed in the frequency, rather than the size, of the gifts.

The typical pattern of meat distribution was exemplified in the division of a gemsbok shot by a ≠xade hunter, //aūdze, assisted by his partner, Kamadwe. The weights of portions were estimated by me.

Kamadwe took and distributed 27.5 kg:

His own share, for his household	5.0 kg
His married daughter and her husband received	3.0 kg
!āō, and his household, in repayment for a mat, were given	3.5 kg
N//haukhwa, Kinin/u, who helped butcher and carry	7.0 kg
!auka, to divide among his people who were visiting the band	9.0 kg

//aūdze took and distributed 42 kg:

His own share, for his household	5.0 kg
His widowed mother received	2.5 kg
His wife's two younger married brothers, each 4.5 kg	9.0 kg
His elder married brother received	5.0 kg
His two married initiation mates, each 4.5 kg	9.0 kg
His married sister	3.0 kg
N//ein/u, the oldest woman in the band, received	2.0 kg
≠xwa:, a bachelor who is always given meat	1.5 kg
Khwakhwe who, with his wife's two brothers, helped carry	5.0 kg

In this division married persons with dependent children were given the largest portions (3 to 5 kg); couples without dependent children had smaller shares (about 2 kg), and the smallest gifts (1 to 2 kg) went to single persons. Of the visiting band, !auka was a close friend of both hunters and was therefore chosen as the representative of his people. Although !āō kept his household well supplied with springhare and small mammals, he was no longer fit enough for regular bow-and-arrow hunting; actually, he was quite an old man to still have dependent children. Without a partner, he seldom killed anything large enough for division and could seldom give gifts of meat. He was, however, a handy craftsman and made artifacts, which he gave to others. This present of meat was in partial return for a winnowing mat given to Kamadwe for his wife. Divorced, ≠xwo: was living with the bachelors at that time. As the nominal hunter of all that I shot, he had given a portion to //āūdze some weeks previously.

Many of the recipients further divided their portions and passed on small amounts of raw meat to others so that, by the evening meal, almost all households had at least something of the kill. (Cooked meat, like gathered food, is not normally shared outside the household.)

The whole carcass of smaller mammals, such as steenbok, duiker, or springhare, is given as a gift. However, raw meat from these animals does pass between households of close kin (e.g., siblings, parents, and independent children) when hunters are incapacitated through age or illness or when their wives are menstruating.

Artifacts. A hut or shelter and the cleared space around it is regarded as the property of the household in that nonmembers do not intrude without express or implied permission. Ownership persists after the household leaves the hut to move to another campsite. For as long as the structure stands and its constituent materials can be used, no outsider may properly take them. (Although the G/wi do not reoccupy old campsites, they do return to the same areas and sometimes dismantle old huts and re-use the materials in building their new huts.) Water and firewood, once taken by household members, are also household property.

All manufactured objects are individually owned. Ownership is transferable by donation, which is the basis of exchange. It is also the means whereby women acquire most of their clothing and utensils, which are made or imported into the band by men. Once the object has been given, transfer of ownership is complete and irrevocable. Husbands' presents to their wives are not recalled when the marriage is dissolved by divorce, and gifts given on exchange expeditions cannot be reclaimed if reciprocation is inadequate. In January 1963 I was shown how irrevocable transfer of ownership is. Two Nharo and a northern G/wi from the ranches came down, following on my tracks, to ≠xade. They demanded that one of the ≠xade men give them all his tanned hides. He gave them some but protested against giving the whole parcel. The G/wi and one of the Nharo caught him and held him over a fire until he agreed to give them the whole lot. They then demanded that he carry the skins back to the ranches for them. He was afraid to travel into what was to him unknown country, so was again suspended over the fire. The local district commissioner accompanied me on this trip and we arrived a few hours after the events had taken place. The northerners had fled on hearing our approach but were apprehended after a two-week search and were brought back to ≠xade for trial by the district commissioner. He ordered that the hides be returned to their owner, the man who had been threatened with fire. The imported judicial process highlighted a conflict of laws; although, according to Roman-Dutch law, the hides could and should have been returned, this was impossible in G/wi eyes. Their former owner refused to accept them and the ≠xade people agreed that the hides were the property of the strangers because "the skins were given to them." Donation, even under this unprecedented duress, was final. In 1965 the victim of the attack asked me to remind the three strangers, should I see them again, "not to forget him." In other words, he maintained his view of the events as being an exchange transaction and expected something in return for the hides.

Ownership, in the sense used here, is diminished to some extent by the obligations that kinship and friendship impose on owners to lend many of their possessions. Avoidance relatives may borrow "general use" artifacts after asking permission, but a joking relative may borrow without the formality of a request. ("General use" objects include utensils for preparing food; tools for working wood, leather, and iron; dancing rattles, and finger pianos.) The owner is not expected to relinquish possessions that he is actually using at the time, but the obligation to lend does intrude upon the degree of control that an individual has over his property.

At the same time, obligatory lending increases the utilization of the possessions by widening the circle of users beyond the household of the owner to the whole range of borrowers. The increased efficiency of utilization allows the band to maintain the same technological level with a smaller total number of tools and utensils than would be required if each household had recourse to only its own possessions. Two or three metalworking kits (anvil, hammer, and, perhaps, files) are sufficient for the whole band's needs. The rate of use of many other general use artifacts varies according to season; pestles and mortars, for instance, are used intensively in every household in melon season but are hardly used at all in the second half of the year until the *Boscia* berries ripen. In the slacker seasons the band can make do with a smaller number of mortars, and it is not necessary for every household to carry this heavy and bulky equipment on each migration between campsites. Instead, some households may leave their less intensively used implements behind, safely stored in the branches of trees until widespread need again arises. In the meantime the burden of migration is slightly lessened. As all possessions must be carried, it is advantageous to minimize the load and maximize the band's mobility, thus allowing a wider choice of campsite and food supply.

Borrowing incurs the obligation of keeping the article in working condition and of carrying it to the next campsite if migration occurs during the currency of the loan. A borrower is also expected to lend his possessions with at least the readiness he shows in borrowing others' goods. The reciprocation of lending rights is essentially generalized, rather than specific to the borrower-lender dyad. It is, however, at least partly a function of relationship, and lending is most intensive between households linked by strong bonds of friendship and kinship, for example, within a clique and between same-sex siblings. (The loan of arrows is, of course, an exception and requires the specific reciprocation described earlier.)

Exchange of services and goods

Within the band

The exchange of services and goods occurs only between members of different households. The rendering of services and the giving of goods within a household is either mandatory or complementary and lacks the element of reciprocity that is essential to exchange.

Within the band, services and goods are exchanged with greatest frequency between men, less frequently between women (who manufacture fewer goods suitable for exchange and possess fewer skills relevant to the rendering of extrahousehold services), and least commonly between individual men and women. There is widespread

237

and frequent cooperation between the men and the women composing the band, for example, in the performance of exorcising and medicinal dances and in the dissemination of information regarding the location and state of food plants and the movements of game animals, but these activities – like intrahousehold services – are complementary rather than reciprocal.

Men, however, have command over subsistence items and other commodities that can be exchanged far beyond the confines of their own bands, extending to allied and other bands and even to other peoples. Meat, being highly perishable, is exchanged only within the gathered band and between visiting and host bands. The same stricture obviously applies to the associated services of assisting in the process of skinning, gutting, and butchering carcasses and carrying meat to camp. Young men provide a subsistence service by fetching water from distant pools, for which they are not directly rewarded.

Exchange is self-perpetuating in that a gift or service not only discharges a previous obligation but also creates a new one. A man gave a spear blade in return for a mortar and pestle, which he had asked another to make for him. This returned the favor of making the mortar and pestle, but, at the same time, the spear blade was regarded as a gift for which a return would eventually have to be made – and so it would go on, ad infinitum. Apart from this type of specified obligation to reciprocate, there are also social zones of generalized reciprocity, comparable with the generality of lending rights referred to previously. Initiation-school mates are required to help one another and to give gifts of food or artifacts when in a position to do so. There is no expectation of specific reciprocation; one mate helps another and expects that a third will help him when his own need arises. Joking relatives are obliged to grant each others' requests for goods and services and to give these priority over normal (i.e., other) exchange obligations. In theory this could unbalance cycles of individual reciprocation, but in practice conflict between the two sets of obligations never occurred and I did not hear of anybody having to refuse to give a gift because it had been reserved for another, that is, joking partners were neither deprived by normal exchange obligations nor did they make burdensome demands. A man is obliged to give periodic gifts to some of his avoidance relatives (e.g., his wife's parents and his sisters) without expectation of direct reciprocation. The return is more or less circuitous; females do not reciprocate, but their husbands make return on their behalf and parents reciprocate favors done for their dependent children.

Exchange beyond the band

Exchanges between members of allied bands camped together are similar to, although less intensive than, the flow within the band. When a household travels alone to an allied band there is invariably a kinship connection; the visitors attach themselves to the clique of their relative and are thus integrated into the exchange system of the host band. This can be seen as incidental exchange, a by-product of visits not expressly made for purpose of exchange.

Expeditions explicitly intended for exchange purpose are undertaken by small parties of men who visit one or more allied bands to dispose of their own goods in order to obtain imports and other commodities not available in their home bands. The visits are of short duration and few services are included in the exchanges. The visitors are given welcoming gifts of meat and occasionally other shares of kills but are expected to feed themselves. They are accorded the usual free hunting and collecting privileges extended to visitors.

The goods taken on these expeditions are the small surpluses that have been accumulated within the home territory. Those remaining at home pass on what they can spare from their stock of such things as reeds for arrow shafts, softened steenbok and duiker skins, tanned hides of larger antelope and leatherwork to the members of the expedition. The goods are received as normal gifts, for which return will be made after the expedition. Such expeditions are the means whereby tobacco, soft-iron rod, wire, trade beads, and iron- and enamel-ware are dispersed across the central Kalahari, passing from band to band to eventually reach even those living in the remotest parts. These imports, originally bought by Bushmen or Bantu at trading stores at Ghanzi, Rakops, Tjihitwa, Kang, and so on, are sold, bartered, or exchanged as gifts, depending on the parties to the transactions, in return for skins, leatherwork, and beadwork, which the storekeepers eventually purchase.

In the remoter areas of the central Kalahari the gift-exchange system of obtaining imports appears to be largely a matter of chance in which the expeditions have little control over the return they receive on their goods. There is no bargaining, and information on what is available in the intended host band is sparse and outdated. However, the initiative lies with the visitors and it is up to them to assess the exchange potential of their hosts and to bestow gifts where the likelihood of satisfactory return is greatest.

To illustrate the working of the system: Two men left ≠xade with parcels of tanned and softened steenbok and duiker hides. At Easter

Pan they stayed with their allies for a few days but could not find anything they wanted, so they gave only a few gifts and went on to Tsxobe, where one had a relative by marriage. They stayed with this man and looked around for the most suitable recipients for their skins. The Tsxobe people had plenty of iron, which the ≠xade men wanted, but no worthwhile amount of tobacco to spare. It was early in the traveling season and the Tsxobe people were planning a visit to Rakops to buy tobacco and other supplies at the trading store. The ≠xade men took advantage of this by giving some of their skins to those who were going to Rakops. The other skins were given to those who had iron to spare. Because the skins were fresh and well prepared, they would fetch good prices at the store. Although intrinsic or commercial value does not determine the magnitude of the return gift in the normal run of exchanges, the visitors obviously recognized the advantage of their particular situation at Tsxobe at that time; they knew also, of course, that this knowledge was shared by their exchange partners. They waited until the Rakops party returned, got their tobacco, and left for home.

On their way, they again called at Easter Pan. There the people begged for tobacco. They could not properly be refused and tobacco was given in return for skins and an iron three-legged pot. There was also a promise of some tobacco when the Easter Pan people obtained another supply. On their return to ≠xade, the two men brought back skins (which they had had before the expedition), some iron and wire, the pot, and not very much tobacco, yet the trip was considered successful. I questioned why they did not cut out the middleman by accompanying the Tsxobe party to Rakops to deal directly with the trader. They rejected this idea; they did not want to travel through strange country and would not want to deal with a stranger. As it was, they had been reluctant to go as far afield as Tsxobe but had been encouraged by the thought of the affinal relative of one of the party and the fact that he could be a link with others in the Tsxobe band with whom they could effect exchanges. I thought that the return to Easter Pan had proved calamitous, which view they also rejected. These were their allies whom they could not fail to visit after having traveled through their country. The exchanges had been equitable – the pot was a good one, the skins could be exchanged with the G!õ:sa, Piper Pans, or other people later in the traveling season, and they were assured of another supply of tobacco after the Easter Pan people got theirs.

On a later occasion a ≠xade man, N/udukhwe, was traveling north with me when we encountered a hunting party from another band.

N/udukhwe inquired after one of their members and was told that he was hunting in another direction and would probably not return for several days. We could not wait for him, so N/udukhwe gave two large handfuls of tobacco to one of the men and asked him to pass it on to the friend. It transpired that this was the return of a gift that N/udukhwe had received while on a visit to this band a year previously. He had been given less than the amount that he gave, but he had passed on the larger amount because, as he explained, he had plenty of tobacco and could get more from me when he needed it. His friend, on the other hand, was with a hunting party and had no other source of supply and would have to share this amount with his colleagues.

Extraband exchange partnerships form a network over the whole central Kalahari, crossing band and tribal boundaries and connecting with the national economy at trading stores, ranches, and cattle posts. Partnerships are initially developed within the kin group but, as the first of the two above-mentioned examples shows, temporarily fan out from this nucleus. Stable partnerships may later develop in this way. The partnerships, which are activated at irregular intervals, provide the means of circulating imports and exportable surpluses, establish and maintain bonds between bands, and allow distant kinsmen and friends to keep in touch through the news carried by men on exchange expeditions.

It is clear that there is a notion of value in the giving of gifts and of the appropriateness of the nature and magnitude of the return. Value appears to be a function of the needs of the recipient weighed against the availability of the commodity. The variability of both these factors precludes stabilization of relativity between commodities and there is no certainty of the precise return that the outlay of a particular gift might bring. However, instability in the "exchange rate" of specific commodities is countered by the stabilizing redundancy inherent in the diversity of both media and channels of exchange and the long-term credit extended to partners. Reverting again to the ≠xade Tsxobe expedition, the exchange value of the skins at Tsxobe in relation to tobacco was protected, first, by the fact that the ≠xade men were prepared to wait unti the Tsxobe returned from Rakops with fresh supplies and, second, by the fact that tobacco was in competition with iron as the medium of exchange for the skins. Conversely, at Easter Pan, the high value that scarcity might have lent to tobacco was depressed by the fact that the ≠xade men were prepared to accept a promise of later gifts of tobacco. The other channels of exchange open to ≠xade (Piper Pans, G!ō:sa, etc.) gave liquidity and value to the skins that the Easter Pan people offered. There had been a surplus at ≠xade, which might have depressed the value of the skins were it not that they could still be exchanged at other outlets in return for commodities that were needed at ≠xade.

Work and allocation of time

An overall seasonal rhythm limits the amount of time available for activities other than gathering and preparing food. The G/wi year begins after winter. At this time the nutritional state and energy level of the people are low. Day temperatures rise into the high forties, and because of the shortage of fluids in the diet, body water levels must be conserved by avoiding excessive losses through perspiration. Physical exertion is consequently restricted to the cooler morning and late afternoon hours, that is, from sunrise to about 10 A.M. and from 4 P.M. or later to sunset – a working day of five or six hours. There is only a narrow range of food plants available, and these are thinly dispersed. Gathering a day's ration therefore takes up most of the working day. (The middle hours are spent resting, lying in the scant shade and sleeping, or trying to keep tolerably cool in a patch of urine-moistened sand.)

As the year advances, conditions are ameliorated by the wet season, after which the climate is less stressful and food plants are more varied and numerous. The midday rest is no longer a survival necessity. Less time and energy are needed to collect a day's food ration, more time can be spared for other activities, and there is a corresponding proliferation of occupations to which people devote that time. This tendency increases with the onset of the wet season, rises to a peak in late summer and autumn, and diminishes as winter deepens.

For most months of the year girls and women are occupied with tasks of household maintenance. Much of the day is spent on food gathering, a task that is usually laborious and must be repeated every day in all but the most favorable seasons. Preparing the gathered food occupies the late afternoon and evening. The burden of migration, that is, packing and transporting household possessions, is borne equally by men and women, but the task of building the hut or shelter at the new camp and setting up the machinery of the household usually requires more work from wives than from their husbands. It is only in autumn that there is an appreciable amount of daytime that women can spare from household tasks; then they are free to remain in camp and engage in the amusements of dancing, musical games, and talk.

After dark the only light is that of the small fires that the G/wi keep constantly burning for warmth, cooking, and to save the labor of starting a new fire. The evening meal lasts until well after dark and is an informal affair with household members eating for short periods and then going off to talk to neighbors or to those in other cliques. Gradually a crowd coalesces about the fire of one or two households. Sitting or sprawled in the sand, the people spend two or three hours in lively talk, discussing the day's events, plans for the future, and other matters of current interest. In this manner most of the information necessary to the conduct and maintenance of band life is circulated and most decisions are taken. It is also a time of amusement when some of the rich store of G/wi folktales are told and enacted. From late January onward, two or three nights of each week are spent in curative and exorcising dancing in which practically all adults participate.

Men are less tied to the routine of household management and their occupations are more varied. The younger men who are still fit for hunting must choose from four mutually exclusive sectors of activity in allocating their time, namely, "make and mend," hunting, travel, and amusement. The amount of time actually allocated follows that order: Work connected with the processing of raw materials and the making and maintaining of tools, weapons, and other artifacts takes up more of men's time than does any other category of work. In general terms, it is the way in which men spend their time when not hunting or traveling. "Make and mend" does not involve constant exertion; work carries on between frequent pauses, and many of those remaining in camp are helpers whose assistance is needed for only short periods at long intervals. I have indicated earlier the approximate average time taken to make various artifacts and to prepare materials. Actually, the tasks usually take longer to complete. One man stops for a smoke and all work around him is halted as the pipe is passed around. Desultory conversation drifts to a topic of general interest and becomes so animated as to demand wholehearted concentration, and work is again interrupted. Hunger may send a man to his hut for a snack and his example sets others off on the same track or out into the veld to gather a few mouthfuls of berries or other plant foods. The men potter, rather than apply themselves steadily to one task after another, but the conversation and interaction play useful parts in communicating information and sustaining harmony.

The individual patterns of hunting are idiosyncratic and are too variable to allow of generalization. Only a small number of men are consistent hunters, going out on every day that weather conditions and the state of their weapons permit. Others follow this pattern for as long as three months and then suddenly change to make and mend occupations for five or six weeks. Although some men switched because they had accumulated a stock of hides and needed to concentrate on tanning and softening them, others had no discernible reason for changing their habits and did so only because they had grown tired of hunting. Others, again, would hunt for only some ten days in a month and then seem to grow enthusiastic and make daily sorties for up to three weeks at a stretch.

At the band level the pattern is no more uniform: In good weather between two and six groups (i.e., pairs or solitary individuals) of hunters are in the field each day for between four and ten hours. This fluctuation bears little relationship to the amount of meat brought into camp, to the number of animals killed, or to the amount of game present in the area.

As men grow older they spend less time on bow-and-arrow hunting. As their strength, eyesight, and wind deteriorate, they rely more for their supply of meat on springhares and on what comes into their snares. Their prey is more localized and the techniques of hunting it are less time-consuming and thus allow the older men to spend a portion of the day on handwork. There is a tendency for them to devote more time to make and mend occupations and to exchange such products with younger and more active hunters for meat. The restricted inventory of artifacts and the forced independence of households during periods of isolation inhibit the development of what might otherwise have become a complementary division of labor by age.

The time spent in travel varies greatly among individuals and from one year to another. If the season is good, half the men in the band will be absent from home at one time or another unless their own band visits, or is host to, an ally, in which case the purposes of travel are met by the visit or the visitors. Individual traveler's periods of absence (i.e., as distinct from the whole band's visit) vary from five days to the whole three months of the traveling season. (Longer absences, stretching to a year or more, are not included in this category of travel, which is made for purposes of exchange, passing on and gathering information, and social interaction.)

Men spend somewhat less time than do women on amusement, dances, and games. However, having the more flexible daily routine, men are able to engage in these pastimes for a larger part of the year, and an hour or two of most days between March and the onset of winter is given over to dances and games in which half or more of the men and boys present in camp participate.

Economic choice

Although it is not a generous habitat, the resources of the central Kalahari are plentiful enough, and the technology of the G/wi is sufficiently versatile to present them with a choice of courses of action and the need to select from these. The alternatives are perceived by the band or household concerned, and decisions are reached in the light of shared values after weighing factors known to all. That G/wi society has mechanisms for resolving differences of opinion as to which is the optimal alternative (see Chapter 4) indicates that equivalent or nearly equivalent alternatives present themselves with some frequency. These hunter-gatherers are not automatons whose daily lives are narrowly dictated by their environment; the latitude that exists in the choice of allocation of valued means to desired ends is such that costs, in terms of social consequences and time and effort, must be assessed in the light of the anticipated rewards of sustenance and social harmony.

The most obvious sector of economic decision making is that which concerns the occupation and utilization of particular areas of territory: the timing and tactics of the seasonally varied patterns and the band's timing and destination of migration to its successive campsites (Figure 16).

In adopting patterns of territorial occupation, the band's decision to split up into individual households and to retreat into isolation is in many years hastened by the occurrence of severe frost, for the suddenly and drastically depleted local supply of plant foods can no longer support joint camps. But however sudden the frost, there is always some warning, and the decision of where each wintering household shall be located is reached before weather conditions can precipitate crisis.

In my observation no wintering site was ever used twice in any 3-year period. Informants confirmed my deduction that annual climatic variation, particularly rainfall, is sufficient to nullify the potential permanent advantages (in terms of vegetational variety or density) that one locality may have over another – last year's optimal site may be incapable of supporting a household this winter. The list of possible sites can, therefore, only be drawn up after the current year's rainfall and floristic states have been reviewed. Discussion begins after the last rains have fallen. Information is passed around, collated and compared, and the potential of various regions is assessed. The requirements of a wintering household are, primarily, a sufficiency of esculent plants to meet their nutritional and fluid needs. Second, there must be trees and shrubs to shelter against the

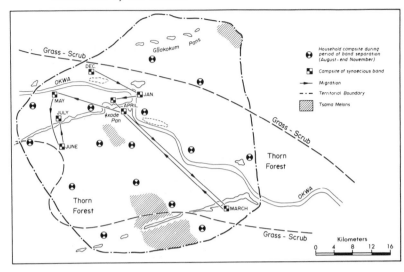

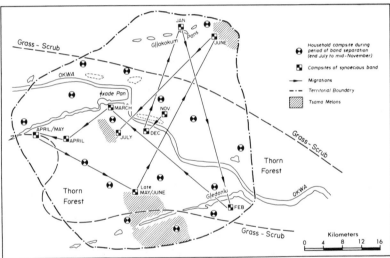

Figure 16. Migration in ≠xade territory in a year with a poor tsama season (*top*) and in a year with a good tsama season (*bottom*).

wind, to shade in the heat of early summer, to provide firewood to cook the staple *Coccinia* tubers (poisonous in the raw state) and other food, and to ward off the bitter cold of winter nights. So that trapping may bring some meat and hides, a population of steenbok and duiker is desirable – and easy to come by, for these small antelope are

246

virtually ubiquitous and, perhaps because of their territoriality, are very evenly distributed. Finally, a wintering household requires sufficient space, not only for gathering and trapping but also for hunting. It is in this regard, particularly, that boundaries of household ranges are decided in some detail. A wintering hunter covers about 75 km² during the season of isolation. This approaches the modal density of territorial population in the central Kalahari and would, therefore, entail a nearly even distribution of households over the band's territory were it not that etiquette demands that one hunt *away* from a neighbor. (Because little hunting is done during isolation, the signs of a neighbor's activity are there to see. If recent footprints are encountered, the hunter changes direction to avoid his neighbor; privacy is inviolate and the household range is not intruded upon while in use.) An overlap of ranges is thus made possible and their distribution within the territory is not even.

The allocation of ranges is largely a matter of each household indicating its intended wintering site and range. Competition is normally precluded by the small size of the band's population. However, when ≠xade band became too large, a shortage of wintering ranges in drought years (perceived as a sympton of overpopulation) was overcome by doubling-up some of the households. Hunting partners joined together and some independent couples joined the parents of one or the other of the spouses. As far as I could ascertain this solution originated in the households that adopted it, but its necessity was unanimously regretted.

When sudden frost does not hasten dispersal, the decision to leave the joint camp is made by individual households and the process of dispersal lasts several weeks. If the band camp is to be moved to a distant part of the territory, those households whose winter ranges lie near the current camp will usually separate at or near the time of the move and save themselves the trek back. Otherwise, the decision to separate is generally prompted by the wife who complains that food is scarce and gathering too difficult. The last to depart are usually the younger, more sociable couples who wring out the longest possible period of shared company. Children are upset at parting from their friends, and this consideration keeps some families in the joint camp for longer than the parents care for.

Before separating, the band decides on the location of its first reunited camp. This decision is in the form of a contingency plan, dependent upon such factors as the severity of the coming winter and the first months of summer, the prospects for rain and the anticipated date of the first rains. A floristic forecast is made for each foreseeable set of circumstances: Should there be a severe winter with sharp frosts but a not-too-severe early summer then such-and-such locality would be the best; should the early months of summer be very bad, the date of rejoining should be postponed until the time of the shepherd's bush berries, and such-and-such would be better because it is located near a grove of these trees and close to large patches of *Ammocharis* lilies (the juice of which is spread over the body to keep cool). In their choice of early campsites the band will avoid the country around waterholes if good rains are expected because these areas will probably be heavily exploited during the wet season. In short, a location is agreed upon for each of several probable sets of circumstances.

The decision to move to the selected joint campsite is reached by each household after a comparison of the state of resources in the wintering range with the probable state of those around the intended site. The household must also decide whether moving at a particular time is worth the effort and whether the inevitable discomfort of early summer joint camps is not too high a price to pay for the pleasure of others' company. The balance is always weighted in favor of moving by the appearance of tracks or other indications of another household's trek to the new site. The band is usually at or very near full strength within two weeks of the arrival of the first household at the joint site.

Once the band has reverted to synoecised occupation the destination of migration is decided largely in terms of the tactics of exploiting the most rewarding food supply at the lowest cost. The evaluation of reward varies with the amount of food available at the new site and with the G/wi scale of preferences among foods. In nearly all migrations the supply of plant foods is the main factor to be assessed. The supply should not only be plentiful but should also be varied and include a sufficiency of the favorite species currently available. Other migrations are directed by the decision to exploit other types of food, for example, "eland" caterpillars. In the wet season the state of waterholes is a factor that must be considered, along with the question of food supplies. A full waterhole, in itself, is not sufficient attraction; there must also be food. But given sufficient food, a waterhole site is usually preferred to a waterless site with more food. Hunting prospects seldom affect the choice of site. When such prospects are a factor, the attraction is the population of sedentary small mammals and the presence of large game. The mobility of both hunters and their larger prey makes it pointless to locate a camp near a herd of antelope. When other rewards are nearly equal, the presence of a supply of poison grubs, grasses suitable for arrow shafts, and wood for making particular artifacts may be a deciding factor.

In assessing the cost of exploiting resources, the band considers not only its next move but the whole series of migrations in the foreseeable future (that is, up to and including winter dispersal). The aim is not to plot the coming season's itinerary in detail but to work out a series of moves that will permit the band the widest choice of subsequent sites. It is a contingency plan with the character of a Markov chain: The alternatives that present themselves are influenced by the moves that have previously been made, for example, the exploitation of an area early in the year will diminish its utility in later months.

The large amount of information and the high rate of information exchange required for calculating moves in this logistic chess are maintained by the participation of all adults in the decision-making process. Among the range of alternatives, only the one that is actually chosen and acted upon is ever put to practical test, that is, there can only be informed conjecture about how other campsites may have compared with the one actually occupied. This could be a divisive factor in times of adversity were it not for the G/wi style of consensus politics and its inhibition of recrimination by factions originally opposed to the decision that was eventually adopted.

Nash (1966:3) defines the term "economy" as "the concrete set of activities and organizations through which a society patterns the flow of goods and services." Sahlins (1965) views the economy of a small-scale society as a collection of processes through which that society provisions itself. In the context of a socioecological analysis, this economy provisions G/wi society by providing systems through which environmental resources are allocated, allocated resources are managed, and goods and services are distributed. To function effectively in provisioning society, the G/wi economy must accommodate the fluctuations in the state of resources and in the pressures that the environment imposes on the society. The economy requires mechanisms that will correct for these perturbations but not disturb its overall stability as a system. In its operation, the system tends to maintain an equality of access to resources and to equalize the distribution of commodities and services despite seasonal and local environmental differences and individual differences in strength and skill. The character of the economy is facilitative and egalitarian, rather than competitive and stratified.

The basis of resource allocation lies in the G/wi doctrine that the deity is the owner of the world and that he ordained that man, as one of his creatures, may use the resources of the world to meet the needs of survival in reasonable comfort. Private ownership and control of the land and its unexploited resources are thus precluded. The doctrine also implies that none may be denied access to resources.

Adherence to this doctrine is impossible if competition is not controlled. Control is facilitated by maintaining a surplus of resources in relation to the needs of the human population. The means of control on a population-wide basis lies in the strategy of locating communities in areas containing a resource nexus and in allocating to each community exclusive rights of occupation and exploitation of resources. Although the community per se is attached to its territory, there is a looser bond between individual and territory. The band is an open, although finite, community, and although changes of band membership are not very frequent, they are not looked upon as abnormal or deserving of opprobrium. Community perception of overpopulation is sensitive and the symptoms are recognized before a crisis develops (see, for instance, the reaction of the band that experienced a shortage of winter ranges, described earlier). The high fusion valency and low fission inertia of G/wi social organization, the absence of firm structural bonds between a household (or individual) and any particular band, and the permissive attitude toward changes of membership facilitate movement of population in response to scarcity in an overpopulated territory, either to establish a new band or to emigrate to bands whose territories have sufficient resources. Any significant differential in resources between bands can be corrected by redistributing population to counter the threat of interband competition contained in the differential.

Resources, then, are allocated among bands, at which level the processes of both allocation and management are contained in the band strategies of territorial occupation.

As a corporate group, the band is both the holder and the administrator of territorial rights. Its egalitarian structure and the beliefs that cast man in the role of N!adima's sufferance-tenant function to confer on all members of the band full and equal rights to its territorial resources. The residential group exploits and consumes the resources in a series of cooperative, or integrated, coordinated operations of the strategy of joint migration and seasonal dispersal.

Migration to the next campsite is timed to occur before the resources of the last become depleted to the stage at which interhousehold competition might arise to threaten cooperation and dislocate coordination. Dispersal of households into isolation occurs before scant supplies enforce frequent and costly shifts of camp. The strategy of timely migration followed by a seasonal reduction in the size of the residential group to match the localized population density to the reduced subsistence density aims at maintaining the most favorable possible balance between supply, demand, and the cost of exploitation. Community coordination in exploitation processes and the savings function of the exchange network (see the following discussion) serve to spread the energy cost of exploitation evenly among the households of the band.

I have argued (see Chapter 4) that the G/wi ideal is the enjoyment of the widest possible circle of harmonious relationships. The alternation of synoecised and separation phases of band life is an ingenious device to accommodate within the ambit of one community a population of the size that can be supported during the seasons of higher environmental productivity. In other words, it is a strategy to maximize band size and indicates the value the G/wi place on having as large a residential group as possible. Translation of this strategy into the tactics of resource management occupies the band with intricate economizing processes that seek the optimal compromise between the conflicting aims of (1) exploiting the requisite amount and variety of resources (2) at the least cost of time and energy in order (3) to retain intact for as long as possible (4) the largest residential group that can be supported by the available resources.

The strategy of resource management restricts band members' rights of exploitation by subjecting them to the discipline of the agreed tactics and the coordination of activity necessary to its implementation. The outcome of a successful program is to ensure equality of access to the best available supply of resources and the continuation of that supply. The need to accumulate surpluses of food does not, therefore, arise. (G/wi technology, in any case, lacks the means of preserving foods, and storage capacity is limited by the relative frequency of migration.) As far as gathered foods are concerned, the equality of access and exploitation cost amount to virtual equality of distribution in relation to each household's needs.

The exchange network functions partly as a savings mechanism in respect to meat and other commodities (excepting gathered food) and also to equalize the distribution of the commodities included among exchange media. The liquidity of media and the opportunities for transactions are enhanced by the openness (i.e., nonexclusiveness) of the network. There is a wide range of goods and services that can be exchanged one for another, and there is a large proportion of the population to whom these goods and services can be given or rendered. When a large antelope is killed, the surplus meat is dispersed among the households via the exchange network, removing the hunter's surplus and tending to equalize distribution. Those who do not hunt are able to reciprocate (in either generalized or balanced modes) with media other than meat. Other hunters may use the surplus from their own kills at a later date, in which case the savings function of the network operates: The first hunter feeds meat into the network and receives another hunter's surplus in return at some future date.

Sahlins's (1965) analysis of primitive exchange is usefully applied to the G/wi case. The savings function of the exchange network is the operation that Sahlins terms "pooling." However, he distinguishes two types of economic transaction: "First, those 'vice versa' movements between two parties known familiarly as 'reciprocity' . . . The second, centralised movements: collection from members of a group, often under one hand, and redivision within the group" (Sahlins, 1965:141). The G/wi savings mechanism is an example of the second type, which occurs as a series of transactions of the first type. Storage occurs in the band network as a whole and not "under one hand." In Sahlins's words, "pooling is an organisation of reciprocities, a system of reciprocities" (1965:141). The social system in general and the kinship system and its elaborations in particular provide the organizational framework within which the "system of reciprocities" operates. There is, therefore, the measure of coherence required to ensure a return from that which the individual feeds into the network. The wide scope of the network, both as to participants and exchange media, induces sufficient redundancy in the system to give it stability in the face of fluctuations in the contributions of individual participants and in the state of supplies of particular commodities.

Sahlins (1965) pleads the analytic wisdom of separating pooling from the two-party transaction by which it is here effected. In his scheme of reciprocities (145 ff.), he distinguishes generalized, balanced, and negative modes of interparty transaction and predicts an association between socioeconomic circumstances and particular types of reciprocity. He examines kinship distance, kinship rank, wealth differential, and the character of exchange media in relation to the dependent variable of reciprocity type. He hypothesizes:

(1) a progression away from generalised reciprocity with increasing kinship distance;

(2) a tendency towards generalised reciprocity between ranks, often accompanied by pooling centricity;

(3) a tendency towards generalised reciprocity where there is substantial wealth differentiation or where general shortages occur (with a rider that the trend may be towards the opposite pole of reciprocity in some instances);

(4) insulation of the food circuit from the main exchange network with a tendency to restrict staple food exchanges to close kin or friends and then in the mode of generalised reciprocity.

His fourth point is hedged with qualification: first, that it is concerned with staple foodstuffs and, second, that these foods are given in expression of sociability, which Sahlins does not see as falling within the economic ambit. It seems to me that such exclusion is contradicted both by his definition of "economy" and, to some extent, by his subsequent (145 ff.) discussion of balanced reciprocity in such contexts as friendship, expression of corporate status, alliances, and so on.

With some reservations, Sahlins's hypotheses are confirmed by the G/wi case. Reciprocity is, indeed, more frequently generalized among close, rather than distant, kin. It is unequivocally generalized between, say, parents and children where reciprocation is not so much back to the givers as onward to the next generation. Grown children return less to their aged parents than they received from them during the years of dependence but give to their own children what they received from their parents. Beyond this range, extending to independent siblings, joking partners, and even to initiation mates, the reciprocation of exchanges is not balanced between giver and recipient by a matching return by the latter. However, within the circle of generalized reciprocity, there is a tendency to overall balance between what an individual gives and what he eventually receives from whomever in the circle. This is not to deny Sahlins but to lend point to his statement that reciprocity is "a continuum of forms" (1965:144).

Furthermore, generalized and balanced reciprocity are not mutually exclusive. As Sahlins indicated (1965:152), certain modes are dominant in particular sectors. In the G/wi case the practice of generalized reciprocity between all but very close kin does not preclude concomitant exchanges in the balanced mode.

Kinship ranking is virtually restricted to the parent-child relationship and to that between a wife's parents and their son-in-law. In the former case the flow of goods is downward. The latter case is somewhat anomalous: The preponderance of goods flows from the son-in-law to the wife's parents. Although it is convenient to refer to this

253

phenomenon as groom service, suggesting that the flow of goods represents the marriage prestation and, hence, balanced reciprocity, the arrangement is not regarded as contractual by G/wi informants. In their view the husband makes his gifts to demonstrate his ability as a hunter and provider and also as tokens of his respect for his wife's parents. The extent and exact nature of the gifts are not specified beyond the notion that a son-in-law should make a present of meat, skins, or other commodities at least once a year. It seems, therefore, that this has more the character of generalized reciprocity up the scale of kinship ranking, rather than the balanced reciprocity of the marriage prestation – giving meat and so on, in return for the wife whom he received from her parents.

The third sector of generalized reciprocity, where significant wealth differential exists, is only applicable to the G/wi case in the context of interband alliances functioning as "drought relief." I refer here to the practice of joint migration into the territory of an allied band that has enjoyed a good season when one band has suffered localized drought in its own territory. Although this is certainly not balanced reciprocity in the sense that the migrating band gives its hosts anything in direct return for the succor received, things even out in time – or at least they are supposed to in the vernacular model – when the migrants find themselves hosts to other drought-stricken bands in the future. This, then, is comparable with the generalized-balanced reciprocity that occurs in the circle of initiation mates, joking partners, and others.

Sahlins's contention that staple foods are insulated from the rest of the exchange network is, of course, true of gathered foods in the G/wi case. It is not valid for the meat of larger game. Sahlins implies that food giving for "sociability" purposes is somehow different from other exchanges and ought, therefore, to constitute a separate category. It is my contention that media other than food are inserted into, and a large part of the exchange network is at times occupied with, transactions that have sociability as their main purpose, but that these, too, have economic significance. These transactions exhibit no formal or other distinctive characteristics. They can only be recognized after examination of the recipients' pragmatic need for that which is given. I can see only confusion coming from their being treated as a special case or series of special cases. As I have said, Sahlins himself seems undecided about the singularity of "sociability" exchanges, as he unequivocally includes in normal balanced reciprocity those exchanges that mark alliances, friendship, and so on (1965:176 ff.) and his view would be difficult to sustain in a situation of competitive sociability in a sort of inverted negative reciprocity. It seems clear that the G/wi consider meat an exchange medium that may be used with all propriety in exchanges characterized by generalized as well as balanced reciprocity.

Sahlins's definition of balanced reciprocity is perhaps unduly restrictive: "transactions which stipulate returns of commensurate worth or utility within a finite and narrow period" (1965:148). The G/wi notion of a recipient's needs as a criterion of evaluation and the elasticity of the time scale of reciprocation might exclude all exchanges from this category. In balanced-reciprocity exchanges "the parties confront each other as distinct economic and social interests" (1965:148). To me this implies an opposition that belongs more properly to the negative mode of reciprocity. I have, instead, relied on Sahlins's "pragmatic test of balanced reciprocity," which is "an inability to tolerate one-way flows; the relations between people are disrupted by a failure to reciprocate within limited time and equivalence leeways" (1965:148), for the meaning of the term.

Potentially, the G/wi exchange network covers the whole range of an individual's acquaintances – his or her social circle. Within this area, exchanges are in either generalized or balanced modes. Negative reciprocity is antithetical to the G/wi view of acceptable interaction and is not normal practice. Where it does happen that one person is disadvantaged by another's actions (e.g., flagrant adultery, rape), the community acts to discourage such disruptive behavior. Although reparation is not effected, there are positive cooperative efforts made to repair emotional damage. Accidental or malicious damage to property by dependent children or by a wife does call for reparation by the vicariously liable parent or husband.

The empirical test that the G/wi case provides of Sahlins's predictions yields only a narrow range of instances that are congruent with his hypotheses. This illustrates the openness and lack of exclusiveness of the network and the small extent to which the economic system prescribes the vectors and media of exchange.

In this open, flexible network the importance of the recipient's needs as a factor in evaluating prestations functions to inhibit the accumulation of wealth by an individual. As a general rule, the more he has, the less he needs and the less will be the relative value of what he receives. An informant epitomized this aspect of the economy: "You do not give a large piece of meat to a man when his pot is already full." The standard of values inclines the gradient from the "haves" toward the "have-nots," and the resultant flow tends to equalize the distribution of goods. Flow is facilitated by the versatility of exchange media, the openness of the network, and the limited circle of generalized reciprocity (see Sahlins, 1965:165–167).

In discussing the technologies of what Cottrell (1955) has termed low-energy societies, Nash (1966:22) concludes that "simple technologies and minor labor specialization give peasant and primitive life a sort of precariousness absent from industrial societies." He argues that the "productive output" of a society is limited by the state of development of its technology and organizational complexity (e.g., "the width of the division of labor," 1966:22). Now it is true that G/wi material technology is simple; the processes by which raw material is transformed into finished products are relatively brief, uncomplicated, and require small investments of energy. In obedience to the principle of conservation, the energy return from artifacts is of a matching low order. In other words, the material technology neither appreciably enhances the physical power of band members nor does it furnish a means of storing energy for any length of time. Energy passes across the G/wi exchange network mainly in the form of foods and physical assistance (services). A high proportion of energy taken into the economic system is expended in the maintenance of the system (notably in the metabolism of its manpower), and a given quantum can only travel for a short distance before being consumed in maintenance processes. The energetics of the economy, therefore, is characterized by short, predominantly linear energy chains of low conductivity. By contrast, a more developed economy's energetics is such that its energy chains are longer, complex, and weblike with higher conductivity. (For instance, pastoralist-cultivators can store energy for fairly long periods in the form of grain cereals or draft animals, both of which can pass across the exchange network in a complex series of moves and transformations as cattle are exchanged for grain and both for labor and so on. The energetics of an industrial economy is even more complicated, with fuels, machines, money, credit, and so on, adding further possible permutations and extensions of energy chains.)

The relative complexity of energy chains seems to me to offer a clearer measure of the state of development of an economy than does Nash's notion of "gross product" (1966:22) in which there is stipulated no common denominator in terms of which comparison can be made. Nash himself (1966:22) mentions the use of M. G. Smith's (1955) concept of a subsistence/energy ratio to describe an economy. Smith's ratio provides an indication of the length and conductivity of energy chains; the higher the ratio, the shorter the chain and the lower its conductivity. Unlike Nash's "gross product," a synthesis of

Smith's concept and that of the energetics of an economy does take into account the systems-maintenance function of exchange.

As Levi-Strauss (1969:52–60) and others have pointed out, economic needs do not constitute the only motivation for giving goods and services. What Sahlins (1965) terms "sociability" is also an important motive. Much of the traffic in the G/wi exchange network has little to do with overt economic needs but is concerned with expressing and maintaining relationships between transacting parties. What Mauss (1954) and Levi-Strauss (1969:52–60) are perhaps saying is that the exchange of prestations in human societies is analogous with the flock calls of gregarious avian species (see Welty, 1964:191–193; Thomson, 1964:310; Eibl-Eibesfeldt, 1970:116 ff.) – "releasers" of social behavior. (If this is indeed Mauss's and Levi-Strauss's thesis, then "prestations" must be intrepreted as covering the full spectrum of forms of coherent and appropriate social feedback.) Certainly the "sociability" aspect of exchange increases the number and frequency of transactions in the network, stabilizing and strengthening it and reinforcing its efficacy as a means of distributing goods and services. As Dalton and others quoted by him have pointed out, "the primitive economy . . . is 'embedded' in other community relationships and is not composed of associations separate from these" (1969:73). There is a functional interdependence between the maintenance of social relationships and the maintenance of the economy. Sahlins's (1965) distinction between "economic" and "sociability" is of analytical and heuristic convenience: Nash's (1966) concern for the precariousness of primitive economies seems to concentrate on the relative mass of materials in circulation and not on the stability and adaptability of the economic systems themselves. The short energy-chain economy does, indeed, have a propensity for instability in the sense that it lacks the bolstering effects that their greater size, complexity, and redundancy have on more developed economies. However, the G/wi economy derives its adaptability and, hence, overall stability from the lability that is a consequence of the instability of the short energy chains in their setting of a social organization that has both the necessity and means to accommodate periodic fission and fusion when the band separates for the winter and reunites for summer. The short energy-chain economy can coherently expand and contract in response to changes in the extent of the interacting social field and the size of the exchange network. In other words, the features that Nash (1966) sees as making for precariousness are precisely those that contribute to the flexibility and versatility of the G/wi economy in the face of environmental and social fluctuations.

6

Socioecology of the G/wi

Dansereau (1957:257) formulated what he terms "the law of the inoptimum": "No species encounters in any given habitat the optimum conditions for all its functions." In greater or lesser measure, every habitat is hostile to the populations of organisms that inhabit it. The environment exerts pressures on the populations, inhibiting the realization of their biotic potential for exponential expansion. Pressures may be slight, in which case expansion is rapid, or of so great a magnitude as to exterminate the population.

Surviving populations are those that are able to meet the environmental pressures bearing upon them. A population does this by exploiting the environmental resources available to it as shelter, sources of energy and information to create and maintain life-support systems, and the basis of artifacts to supplement the life-support systems. The several species populations in a given habitat are interrelated and interact, each surviving by exploiting or cooperating with others and itself being exploited and restrained by competition. Such an assemblage of populations, which together form a functional system of complementary relationships that regulate the transfer and circulation of energy and materials, constitutes and ecological community. An ecological community and its abiotic physico-chemical environment constitutes an ecosystem. The abiotic components of the ecosystem include the basic inorganic elements and compounds, an array of organic compounds, and such physical factors and gradients as moisture, winds, currents, and solar radiation (see Allee et al., 1949; Odum, 1971).

Within the ecosystem the interspecies relationships are influenced by intraspecies relationships; in other words, the way in which a population is itself organized affects and is affected by its interaction with other populations (Odum, 1971:140–150).

The ecosystem is conceptualized as an open system with inputs, throughputs, and the export of energy, matter, and information involving relationships and interaction with other open systems, including other ecosystems (Margalef, 1968:1–25; Odum, 1971:8–85). As a universe of inquiry it is a conveniently flexible unit that can be

258

expanded to global scale or reduced to the confines of the anatomy of a single individual. Also, analysis can coherently proceed from the study of a particular ecosystem to the wider perspective that embraces groups of interrelated ecosystems of a larger universe.

These generalizations are also valid for those ecosystems that include human populations. In the study of human populations the focus of socio-ecological inquiry is the manner and means of their survival in the context of their ecosystems. This includes man's means of self-organization to achieve and sustain adaptation to environmental pressures and his perception of, and role in, the larger system in which his sociocultural creations operate, namely, the complex of interacting components (of which he is one) regulating themselves and one another in a series of ordered mutual checks and balances.

The overall socioecosystem model is represented in simplified form by Figure 17. The squared slab represents the "scientific" view of the environment, that is, my own understanding of it in accor-

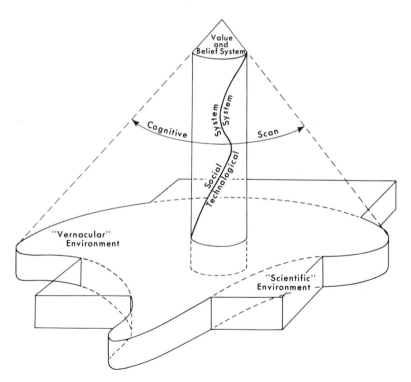

Figure 17. Model of a socioecosystem.

dance with the tenets of European-style natural history. This is what is described in Chapter 2, "The Habitat." The inlaid, extruding slab of irregular shape is the same environment conceptualized by G/wi natural historians (that is, it represents my understanding of their view of it as described in Chapter 3). The sociocultural system is the focus of closer description and analysis and is accorded prominence. It has three components: the economic, social, and cosmological systems. The last includes the values and beliefs of the G/wi. These are not regarded here as independent components but as parts of the whole and can be identified and analyzed as interdependent subsystems. The cognitive scan embraces the whole of the "vernacular" environment and sociocultural system but not the "scientific" environment. The disconformity represents the differential between G/wi knowledge and understanding and that of Western natural historians.

Ecology of the central Kalahari

Characteristics of the ecosystem

In the central Kalahari the change from dry season to wet is marked by an enormous increase in the amount of living material present in the environment and by a corresponding rise in the activity of plant and animal life. During winter and early summer, the ephemeral plant species survive only as inactive seeds, and nearly all of the remaining vegetation consists of deciduous trees and shrubs or grasses, which are also dormant until the wet season or some weeks before its start. The only plants that are active in the dry season are a handful of succulents (e.g., aloes and stapeliads).

Some arthropods hibernate and in the life-cycles of others, like those of ephemeral plants, diapause coincides with the dry season. Many insects, however, notably the social species (e.g., termites, ants, honeybees), do neither. They spend most of their time inside the colonies with their activity greatly reduced (e.g., reproduction is suspended), feeding on what they have stored up during the wet season and only emerging for brief periods when favorable opportunity presents itself for maintenance tasks (e.g., defecation, repair of the nest).

Of the vertebrates, a few hibernate (e.g., amphibians and reptiles) and many migrate out of the central Kalahari. As I have mentioned, about half the bird species are migrants and many of these form large flocks and represent more than this fraction of the total number of the avian population. Similarly, the great number of large antelope that move out for the dry season constitute a significant proportion of the total herbivore population.

The dry season, then, is characterized by a lower level of biomass, a decline in the diversity of species, and a greatly reduced scale of activity. It follows that there is a matching reduction in interactions and relationships within and among species. In other words, the ecology of the central Kalahari undergoes a process of simplification in the dry season.

In the wet season, by contrast, the seeds of ephemeral plants germinate and rapidly cover the bare sand. Their shade not only provides a cooler environment at this level but, by cutting reflection from the sand, also ameliorates the ambient temperature for some meters above. The ground-layer vegetation provides food in the form of foliage, nectar, fruits, and seeds for a host of vertebrate and invertebrate herbivores, populations that are now much greater and more diverse than in the dry months. Higher vegetational strata are changed by refoliation of the trees and shrubs, adding to the shade and fodder for herbivores and also providing visual and climatic shelter for arboreal fauna. Increased atmospheric humidity, amelioration of daytime air temperatures under and around vegetation, and the warmer nights of summer are all less stressful than dry-season conditions and, with the wealth of plant and animal food now present in the habitat, permit a much longer period of activity during the 24 hours. Many animal species reproduce in this season or are engaged in rearing their young, so their range of activities and rates of food consumption are increased.

In the wet season the amount of biomass has increased and a greater number and variety of plants and animals interact for a longer period of the day in a greater diversity of relationships between and within species. In this way the ecology has become more complex.

It is plain to see that this is an ordered ecosystem. Were it not orderd, the G/wi would perish, for they, like the rest of mankind, have not the capacity to adapt to chaos. Their continued presence and the persistence over the period of observation of the same species of fauna and flora demonstrate the overall stability of the ecosystem. Paradoxically, the wet season elaboration of inter- and intraspecies relationships is not accompanied by a matching magnitude of diversification of the spatial structure of the ecosystem.

(I deliberately exclude the microenvironment from this discussion, not because it is unimportant, but because my argument is more easily explained by focusing on the larger-scale phenomena. I believe that my argument holds good for the more elaborately structured microenvironment as well.)

The spatial structure is essentially simple. Competitive exclusion limits vertical stratification to one or two levels over most areas. Few shrubs grow under or close to trees and there is less ground-layer vegetation under either than is found in the open. The overall density of perennial species is low (average coverage varies from 5 to 25 percent), and it is only among ephemerals that horizontal diversity is marked. Few of these latter grow taller than 1 meter and most are of somewhat shorter habit. Competition for moisture and other nutrients is so intense that a ground-layer plant seldom stands in the shade of any other, even though ground coverage is 100 percent in many areas in good years.

The central Kalahari is quartered by the fossil remnants of its main drainages, the Okwa, Merran, and Deception valleys. Although each of the four zones is distinctive in its way, variation within each is of modest extent and gradual, except along the drainages and their tributaries and around pans. Elsewhere, on the sandveld, grassland is dotted by low shrubs and occasional spinneys of stunted trees and merges slowly into areas of low forest. A consequence of this relatively simple vertical and horizontal structure is the absence of a great diversity of spatial location available to animals for development of distinctive niches. Tree-nesting birds, for instance, must accept as building sites a rather narrow range of heights above ground and should not be too specialized in their choice of food, which is available in the immediate vicinity of the nest. Woodpeckers, which spend a tiring fortnight laboriously hammering out nests in tree trunks or larger branches, are likely to find that hornbills, owls, bats, snakes, or bees have moved in a year later. The diets of herbivorous mammals, as reported by Leistner (1967:94–98) and Smithers (1971), indicate that many plants are grazed or browsed by several different herbivores, although no two species have identical diets. In other words, a large measure of overlap exists among the niches of much of the fauna. This overlap is made tolerable (i.e., does not result in a chaotic or unstable ecosystem) by the relative lack of specialization among the animals (i.e., few of them feed off only a narrow band in the vegetation spectrum, are excessively narrow in their specification of shelter or nesting sites, etc.), the consequently high measure of redundancy (or substitutability of available niche spaces), and the freedom allowed them for horizontal separation.

Hypothetically, if these populations were crowded into a restricted space, the overlap in their niches would develop into intolerable competition for the limited resources. Given but one tree, it would either be so riddled with holes as to cause its collapse, or a woodpecker, owl, bat, or whatever would find that there was no room at the inn and do without shelter for her accouchement. But the central Kalahari has enough trees. The poor woodpeckers may be forced to go through the annual business of bashing out a home for their young because last year's address has been usurped and they may have to settle for an entrance only 2 meters above the ground instead of 3 or 4 meters, but they are never short of trees. Antelope must use their mobility to space themselves from others in order to find food, and because in most years the wet season's growth matches the increased demand occasioned by their migrating back into the central Kalahari, they will find what they need. The uniformity of the environment is such that what they need is as likely to lie in one direction as in another.

This elastic ecology accommodates the annual increase in biomass and the complexity of relationships by filling empty space. Much of the ground, left bare toward the end of the dry season, is soon covered by ephemeral and other forbs. The refoliation of deciduous shrubs and trees fills previously empty space among their branches. Herbivores and carnivores hatch, emerge from hibernation, or migrate back into this country. The increase in consumption that they create is met by increased subsistence space and subsistence time. When the land dries and dies and populations contract in the tightening grip of winter, subsistence space and time also shrink. The ecosystem is, in this respect, elastic, like a balloon that can be repeatedly blown up and let down; its surface expands in two dimensions to accommodate increasing three-dimensional volume and then shrinks again, but the balloon retains its structural integrity, its spherical shape. The ecosystem expands subsistence space and the number of niche opportunities but retains much the same, rather simple spatial structure in which there is no great increase in the variety of niche spaces.

Pans and valleys, as I have said, present a sharper ecological contrast than is found in the sandveld around them. The calcareous soils forming the hard, relatively impervious floors of the pans are thought to have been laid down in a pluvial epoch of Kalahari history some ten thousand years ago when rivers flowed and pans were shallow lakes (Lancaster, 1974:168). Today these soils hold pools of water in the wet season and support a more or less specialized flora that provides a living for some surprisingly specialized animals (e.g., amphibians and the Cape terrapin, *Pelomedusa subrufa*, which require standing water for their survival and reproduction). Because more secure burrows can be dug in the harder, more structured tufa than in the sand, the vicinity of the pans is attractive to some fossorial mammals (e.g., ground squirrels), the vegetation is periodically preferred by some herbivores, and the water in the pools attracts nearly all mobile animals. Springbok, in particular, but other antelope, too, favor the pans, gaining some protection from predators on the bare, open surface. In these ways the contrasts between pans and sandveld act as local modifiers of the behavior of species hunted by the G/wi and introudce an element of predictability that is otherwise lacking in the somewhat uniform habitat, where, as I have said, a herd may as well move in one direction as in another (see, for instance, Smithers's (1971:253) discussion of the mobility of hartebeest). Because of the uniformity, redundancy, and randomness, a G/wi band requires space as one category of territorial resources, that is, sheer length and breadth will impose limitation.

It seems likely that there is an interdependence of sorts in the accommodation of migration between the central (and western and southern) Kalahari and the better-watered areas to the northeast. The southwestward movement out of the latter area in summer reduces population pressure during the growing season and allows a buildup of the fodder bank that sustains the migrants and their offspring when they return to these wintering ranges. These more consistently productive areas probably also serve as reservoirs of antelope and other species using this migratory route, for when populations in the central Kalahari are depleted by serious long-term drought, they are replenished from this source. If this link does function in this way, it helps explain the rich diversity of species to be found in this semi-desert.

The position of the G/wi in the ecology

As far as is known, the Bushmen are not a desert ecotype (Tobias, 1964) and have no special, genetically transmitted adaptive advantage over other human populations for survival in this environment. The G/wi sociocultural system must, therefore, meet the same functional needs as do those of other societies (see Aberle et al., 1949–50:100–111) and must serve as the principal means whereby these people use the resources of their environment to meet the pressures that it exerts on them.

As an animal species, man can obtain his nutritional requirements from a wide range of foodstuffs and includes both plants and animals in his diet. As hunters and gatherers, the G/wi occupy positions as predators and herbivores in the trophic network and intrude into the niches of other species in these positions. The stability of their subsistence base is given some protection from disruptive fluctuations in the ecosystem by the relatively large number of species that, to the hunter and gatherer, are equivalent or nearly so, that is, by the catholic diet that their habitat provides, by the coincidence of essential resources in one nexus, and by the devices that the people employ to manipulate the density and location of their own population to match the disposition and amount of resources in the habitat.

Interspecies competition. Each of the esculent species, whether of major or lesser dietary importance, is also eaten by a range of competitors. Birds eat the fruits, antelope and other herbivores browse the leaves and root up the tubers and bulbs. Insects and small mammals add to the competitive pressure. However, with few exceptions, each competitor has a choice of equivalent plants and the pressures that it exerts on any one esculent species is lessened by dilution over the other equivalent portions of the floristic spectrum. The range of plants available to the G/wi is greatest at the time that competitors are most numerous; when fewer species are to be taken, the competitive load declines. The periodicity of esculent plants is such that a range is available in all seasons with a broad overlap in the periods of availability of the species of major importance.

In their hunting the G/wi compete directly with all the mammalian predators, which are themselves in competition with one another. On a short-term, localized basis, the G/wi have a potential advantage in that their competitors normally fear man and run away, rather than compete with him for the same prey animal. However, this advantage is outweighed by the longer diel activity and greater mobility and hunting efficiency of the competing predators.

The effects of gathering and hunting

Variation in the intensity with which the G/wi exploit esculent plants is indicated in Figures 18 through 22. The very difficult conditions of early summer (September) are reflected in Figure 18. The household represented here exploited only four esculent species (*Coccinia rehmannii, C. sessifolia, Raphionacme burkei,* and *Commiphora pyracanthoides*), which yielded both food and moisture. The gross rate of exploitation was then of the order of 0.15 kg/km^2 per day

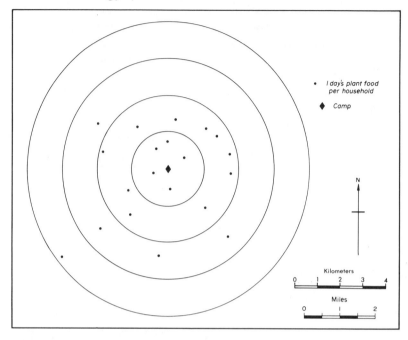

Figure 18. Plant food exploitation during 20 days of subsistence living in September, a lean time of the year.

and most gathering was done within 3.5 km of the camp. In December (Figure 19), with improved conditions, the density of plant resources was sufficient to again support a joint camp. The main food was witgat berries (*Boscia albitrunca*), and because of the location of the fruit-bearing trees, gathering was largely confined to a 90-degree sector of a 6.5 km radius centered just north of the camp. At this time the gross exploitation rate rose to 12.44 kg/km² per day.

In March (Figure 20) there were two dozen esculent species available and the exploitation rate was about the same. There was, however, a good supply of water down on the pan about a kilometer from camp and this compensated for the extra work of gathering in what was otherwise a rather poor area in that year.

In May of the year of the prolific tsama season (Figure 21), another band came to visit, swelling the camp population to 138. It was possible for all to find their plant food within an area approximately 1.5 by 2.5 km in extent. At this time the gross daily exploitation rate averaged 128.95 kg/km². By contrast, in the same month of another year, when the tsama crop was poor (Figure 22), gathering was

266

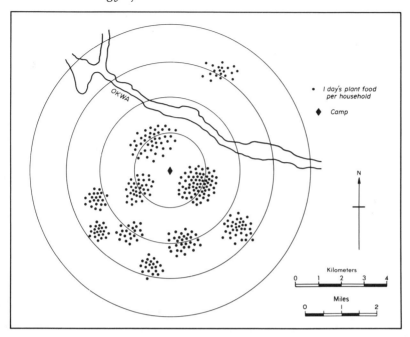

Figure 19. Plant food exploitation during 20 days of subsistence living in December, when conditions were improving.

spread out over a 25-degree segment of an 8 km radius, and the average daily rate was 22.4 kg/km². At this time there were 29 species available. *Grewia* berries, *Vigna* roots, and truffles were most heavily exploited, with *Scilla* bulbs, *Ceropegia* tubers, and cucumbers (*Colocynthis naudinianus*) also making up large portions of the diet.

Gross exploitation rates were calculated on the basis of even gathering throughout the area in which plants were being collected. This was not, of course, the case, but it was not practicable to measure more accurately. The error is fairly uniform and the figures are sufficiently realistic to indicate relative seasonal fruitfulness and density of exploited species.

Exploitation is never so intense as to denude the area. The G/wi believe that N!adima would be angered if there were not enough plants left to ensure regeneration. Migration of the synoecised band is, in any case, timed to occur before food supplies sink to the level at which denudation would occur. At this stage gathering would be excessively costly in time and energy. The efficiency of gathering is rather low in terms of the total amount available in the locality but

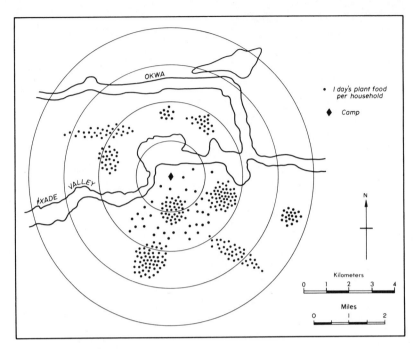

Figure 20. Plant food exploitation during 20 days of subsistence living in March.

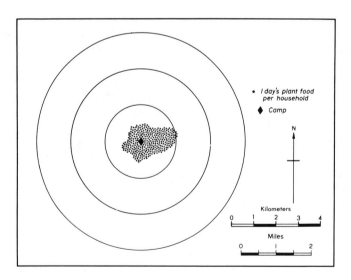

Figure 21. Plant food exploitation during 20 days of subsistence living in May of a year when tsamas were in good supply.

268

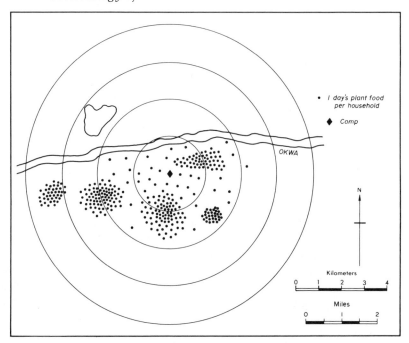

Figure 22. Plant food exploitation during 20 days of subsistence living in May of a year when tsamas were in poor supply.

high in terms of cost-benefit balance. The efficiency of gathering by the isolated, sedentary household on its winter range increases as the family gets to know the area and spends less time and energy on scouting.

A consequence of maintaining a favorable balance between the costs and benefits of gathering is that the load that a household or band imposes on the esculent plant populations is not heavy enough to threaten the stability of floristic communities, which appeared to me to return to normal after the first good rainy season. Ingested seeds that are passed with feces and deposited about the periphery of the camp are rapidly worked into the ground along with the fecal material by the industrious dung beetles. If the seeds remain viable, conditions for germination are highly favorable. It is possible that the occasional prolific stands of *Grewia* sp. and other esculent plants originated in this manner and that the G/wi are unintentional cultivators. Competitive exclusion, however, usually negates this effect after some time.

Contrary to the generalization expressed by Blair Rains and Yalala (1972:50), the custom of annually burning off stretches of grass to encourage the growth of cucurbitaceous ground vines appeared to me to effect only temporary changes in the flora of ≠xade Pan and the neighboring regions. The fires are usually not intense and would, under other circumstances, have the effect opposite to that intended, that is, the growth of grasses would be encouraged, that of ground vines would be suppressed, and shrub encroachment would be inhibited (West, 1955:629 ff.). However, the burned grasses are the first to flush after rain and are usually the only green pasture to be seen after the first, premonitory showers fall. Antelope and other herbivores keenly graze the young shoots and thus set back growth for the whole ensuing season, allowing the vines, which only emerge well into the wet season, to flourish. The grasses reestablish their dominance and the scorched shrubs recover after the first year. The variation of annual migration patterns and the location of winter ranges serve to effect a rotational pattern of burning and it does not normally happen that the same area is burned in successive years. Repeated burning of the same area, which has occurred to the northwest of the game reserve, has produced a marked dominance of unpalatable shrubs, for example, *Dichrostachya cinerea,* and here Blair Rains and Yalala's (1972:50) strictures are certainly valid.

Vegetation is considerably disturbed in the immediate vicinity of G/wi campsites. Trees and shrubs are trimmed to accommodate huts and shelters, ground cover is trodden out, and there is a relatively heavy local concentration of ash, food waste, and other organic material. In 1966 it was still possible to detect the residual effects in camps that had been occupied in early 1959. Disturbance by man is, however, less extensive than the local effects of grazing, browsing, and rooting by some antelope, particularly wildebeest and springbok, which leave stretches of up to 3 acres completely cleared of all but the trees and larger shrubs. Irruptions of rodents, which attain astonishing numbers, also devastate the local flora and leave the area barren for some time afterward by honeycombing the ground with their burrows and preventing the subsurface storage of water needed by plants.

The effect of G/wi hunting on central Kalahari game is clearly something other than the simple predator-prey relationship modeled by the Lotka-Volterra equations. Population cycles do occur, but they are not of the nature predicted by these equations. The populations of all species change in response to such environmental fluctuations as drought and good rains. There are also changes that are not understood and cannot be related to these fluctuations. The "treks" of springbok that occurred in the past amounted to irruptions, and the phenomenon itself, but not its explanation, is well documented (e.g., Cumming, 1863:47, Cronwright-Schreiner, 1925). Other populations (e.g., hartebeest) seem to rise and fall in comparable but less dramatic fashion.

The role of the G/wi hunter is different from that of most predators examined by ecological theory. That this is so is not so much the fault of the theory as that of those who equate hunter-gatherers with other predators and expect them to behave with the simplicity of Gause's protozoan *Dinidium* busily exterminating paramecia in bottles of oatmeal. As Dasmann (1964a:194) correctly emphasizes, hunting is only a "semisubsistence activity" and meat is only a supplement to the diet of southern African hunter-gatherers; they are not the same kind of predator as lions or praying mantids.

Nevertheless, the G/wi hunter's weapons increase his potential killing capacity far beyond that provided by the slight extent of human anatomical specialization in that direction. A bow and poisoned arrow more than compensate for man's lack of strength, speed, and effective fangs and claws and extend his reach far beyond the length of his arm. Our species' ability to transmit knowledge in a sociocultural context and our level of intelligence enable us to learn a large repertoire of search images and to elaborate these by inclusion of data (e.g., spoor recognition, habits of the prey animals, behavior of other species indicative of prey movements), which make it possible to predict the location of prey and thus reduce the restrictions that time and space impose on other terrestial predators. That these search images and ancillary data are applied in a social context by the G/wi has the effect of multiplying the hunter by the number of members in his band who add their information to that which he gathers from himself, further freeing him from spatial and temporal restraints. Potentially, the predation efficiency of the G/wi hunter is very high and he could use that capacity to create a very unstable relationship between human and prey populations in much the same way as did the hunting people of North America who engaged in the fur trade (Martin, 1978). But the G/wi apply their efficiency so as to keep the relationship between them and each prey species at a low level of intensity. As Dasmann noted (1964a:194), they do not have to hunt in order to eat: The dominance of plants in the diet weakens the subsistence tie that intensifies other predator-prey relationships. The table of kills in the preceding chapter illustrates the versatility with which they can switch prey and, consequently, reduce the predation load carried by any one species. The load carried by the whole game population is further lightened by the efficiency with which kills are utilized and by the hunters' great concern not to scare mobile game out of the territory by excessive or careless hunting.

Although it is not intense, there is an ecological relationship between these hunters and their prey. The G/wi are opportunists, but they are businesslike in their opportunism; when a prey species is common in the area, they hunt it in preference to others that are more difficult to find and to approach. In general, they select the prey that will give the greatest reward for the least expenditure of time and energy. In selecting the "cheapest," the locally most easily available, species, they undoubtedly have some effect in damping population peaks. It is, however, no great effect, for, as I explained in Chapter 5, hunting activity is not always proportional to the amount of game in the vicinity. My point is merely that there are differences in the predation load and that, in general, the load tends to be heaviest for the most numerous species.

The efficiency of search, the employment of the combined manpower of the band to feed current intelligence to the hunter, and the wide spectrum of environmental data on which he constantly draws to refine and update the intelligence, focus his attention and activity on a small part of the total hunting range. Search is therefore highly directive and the probability of encounter is heavily biased toward a specific prey category during any one hunting sortie. Apart from consideration of the time-energy cost, other factors that influence the selection of the category of prey are the hunter's age and capability on the day of the hunt. Within the selected category, the object of narrower focus is identified (in the case of larger antelope) by such criteria as age, sex, position in the herd, and "personality." The last two factors are probably interdependent and "personality" is perhaps subject to different readings by different hunters, which would help explain why I found it so confusing and difficult to understand. Whatever the case, this notion of ranking personality types must spread the preference among targets over a wider portion of the prey population than would a choice based only on size (hence, meat yield) and make hunting a less narrow selective pressure on populations of large antelope.

On smaller prey the pressures are distributed differently. The probability of encounter with duiker or steenbok is greatly affected by proximity to the trapper's camp. As he visits his snares several times a day, he can only set them within convenient reach, say, up to 3 km (these visits are not without result, for the trapper combines inspections with foraging trips). Their territoriality makes both these species likely choices for the trapper's highly directional attention, but this is compensated for by their being among the less popular arrow targets. Because a band is dispersed over much of its territory during winter, a considerable pressure is exerted on the duiker and steenbok, which live near household campsites. The other members of these species are subjected to only light pressure. As the locations of household camps are changed each year, the result must be something akin to culling an annually changing stratified sample of the two species' populations. The space depleted by trapping is left undisturbed or nearly so for some years and is therefore available for reoccupation by replacements from the same species.

Hunting pressure on springhares (*Pedetes capensis*) is a more complex matter. The G/wi seldom camp in the open country preferred by springhares. Although hunters will travel moderate distances in search of this prey and will venture into the open country favored by them, the colonies around the pans are likely to receive the most attention, as these are commonly within easy reach of a band camp. The fact that so few are taken in winter reflects the G/wi preference for more heavily bushed country as campsites and wintering ranges. Butynski (1973:209–213) reports much heavier hunting than I observed and mentions a monthly average of six kills per hunter by Bushmen elsewhere in the Kalahari. He does not indicate whether they are from hunter-gatherer bands or are Bushmen attached to some fixed location such as a cattle post or farm. If they are sedentary hunters, the higher figure that he reports is explained by the smaller choice of alternative prey available to these Bushmen and their consequent need to hunt this animal. In addition, because grazing stock and cultivated land create conditions more attractive to springhares, it is quite possible that they are present in greater numbers elsewhere than they are in the nearby hunting area surrounding a G/wi band camp. Butynski is quite right in stating that the desert G/wi could hunt more of these animals, but under the conditions described in this book, they have no reason to.

Tortoises are subject to heavy hunting pressure in their season. As most are picked up by women and men engaged in gathering plant food or collecting firewood, the pattern of hunting is similar to that of foraging, namely, it is carried out within an irregularly shaped area approximately centered on the camp, which is moved every 3 to 6 weeks. As the same foraging ranges may be exploited for days or weeks in succession and then worked over again in the following year (e.g., when camps are located near important waterholes), the same area is hunted repeatedly. However, tortoises are moderately mobile and there is probably a constant summer flow into the hunted areas to replenish the locally depleted population. But they are not so mobile as to necessitate the caution and self-restraint needed when hunting large antelope. One is unlikely to witness herds of tortoises galloping over the horizon because one too many of their number has been popped into some lady's cloak to rest among the roots, tubers, and berries, which the luckless beast will shortly join on the family's menu.

The division of labor, which allows only men to hunt, the theological prohibition against killing more animals than are needed for sustenance, and the restrictions on hunting across territorial boundaries limit the pressures that hunters are able to exert on prey populations. Extrapolating the kill tally given in Table 9 to cover all bands and relating these values to the game censuses that I made indicates that the G/wi hunters' kills add up to a fraction of the mortality caused by the game-control fence on the northern boundary of the game reserve and the depredations of outside poachers (see Silberbauer, 1965:20 ff., 136).

The G/wi diet

The G/wi diet reflects the seasonal fluctuations in the biomass. It is inadequate during early summer, when all complain of hunger and thirst and lose weight. Although none of the classic malnutrition diseases is obvious at this time, the people are more prone to illness, which is probably due to their depressed nutritional status and to the climatic and psychological stresses of the season. A deficiency of B-complex vitamins and vitamin C is apparent from the response to the addition of these to the diet (P. A. Silberbauer, pers. comm.). Because the protein intake is reduced by the seasonal restrictions on hunting, the diet consists largely of carbohydrates and plant fluids.

In other seasons the diet appears to be adequate. Body weight returns to normal and the complaints of hunger and thirst cease. However, even when judged by the somewhat parsimonious standard of the RAF Mk. VI Emergency Ration (which contains 50.85 g protein, 159.14 g fat, and 33.82 g carbohydrate, providing a total of 2946 kcal; Whittingham, 1965:182), the G/wi diet is often deficient in fat. Between November and July the daily meat intake averages about 316 g (25 percent protein) to which must be added the proteins derived from plants, notably tsama seeds, which contain 34 percent protein (Watt and Breyer-Brandwijk, 1962:348). Carbohydrates, as starch and sugars, are present in all the esculent plants. There are, however, few sources of fat. Among the game animals, only eland yield appreciable quantities. Some esculent plants, for example, *Bauhinia esculenta* and tsama, contain up to 40 percent (Watt and Breyer-Brandwijk, 1962), but for most of the year the diet is deficient in this respect by European standards. Salt is of problematic status. It is possible that potassium salts present in *Portulacacae* (e.g., *Talinum* spp.) and in the appreciable quantity of wood ash that is ingested adequately supplement the small amounts of sodium salts contained in the blood and meat components of the diet. At all events, G/wi taste inclines to a far lower salt content in cooked meat than the minimum acceptable to Europeans. Ladell (1957:73) mentions that salt deficiency appears to suppress the subjective thirst threshold. In view of their dependence on plant and animal fluids for water for more than 10 months of the year (including early summer), the low salt intake of the G/wi might be advantageous to them. The low protein levels of early summer (about 15 g per day) are similarly advantageous, as the end products of nitrogen metabolism are highly diuretic. The carbohydrate diet, on the other hand, limits ketosis and diuresis, which is a desirable circumstance when water is scarce (Ladell, 1957:72).

Water

The daily individual fluid intake from plants is estimated to be about 4.5 liters during the hot, dry months. It is possible to survive on much less than this if work is avoided. A fair amount of exertion is possible with 4.5 liters per day, but the sensation of thirst nags constantly, impairing morale, motivation, and the sense of humor. The G/wi habit of resting in the heat of the day must appreciably enhance their ability to tolerate this small intake. Some species of *Raphionacme* contain a principle that has a depressing effect on the central nervous system (Watt and Breyer-Brandwijk, 1962:136); it is possible that the tubers of *R. burkei* on which the G/wi depend for much of their fluid intake in early summer might also have a weak narcotic effect and suppress the subjective sensations of stress. The response of Bushmen to heat is qualitatively and quantitatively similar to the responses of Caucasians and Bantu Negroes (Wyndham, 1956:869–870; Wyndham et al., 1964:885–888). The fluid requirements of the G/wi can be compared, therefore, with those of Europeans and Bantu.

Large antelope killed in the dry season yield large amounts of rumen liquor, which serves as a substitute for water. However, un-

Water is collected from the shrinking waterhole in the middle of the day when game animals are resting and will not be frightened away by the presence of people.

less the kill is made near a camp, the liquor is available only to hunters and their helpers, for it is not usually carried back to camp. This is of little consequence, however, as a reasonably good supply of plant juice is normally available in the season when large antelope are taken. Only small antelope can be hunted in the critical months of September and October, but if by happy accident a household is able to bag a gemsbok or other large antelope, their water problem is solved for the day or so that the rumen contents last.

The maximum shade temperatures of early summer are a little lower than the 52 °C at 18 percent relative humidity that Ladell (1957:44) has calculated to be the upper safe limit. However, because unshaded and sand temperatures reach 60 °C (Wyndham et al., 1964:887) and 72 °C (Cloudsley-Thompson and Chadwick, 1964:15), respectively, thus exceeding this limit, ambient conditions would, in time, cause heat stroke in an unprotected individual. Above 35 °C all heat loss is evaporative (Ladell, 1957:46). The rate of skin-surface evaporation is increased by the strong, dry winds of early summer, and water losses by this route are rapid in the open air. Adolf (quoted in Whittingham, 1965:171) estimated that water loss in excess of 6 percent of body weight – that is, a loss of about 2.75 liters for the average male Bushman – would produce severe debilitation. A loss of 15 to 20 percent (Whittingham, 1965:171) would be fatal. Wyndham and his collaborators (Wyndham et al., 1964:885–888) measured the sweat rate of Bushman subjects and established an average value of 577 ml per hour. A man working between 10 A.M. and 4 P.M. (the resting period) would lose 3.462 liters through sweating and an unknown amount via exhaled air, probably incurring a fluid deficit over the average daily intake of water. Sweat rates during work done in the cooler hours before and after this period drop below the midday values (Ladell, 1957:44). By resting in the shade and cooling themselves with urine-moistened sand or the chewed fiber of *Raphionacme* and *Coccinia* tubers, the G/wi conserve body water and restrict losses to tolerable but not comfortable limits. Many G/wi, by their lassitude, malaise, and atypical irritability manifest what appear to be the prodromal signs of heatstroke, which, although they do not develop into stroke episodes, indicate how slender the margin of survival sometimes becomes under these conditions.

Resource nexus and population spacing

Man's herbivorous and carnivorous diet, which enables him to dodge the pressures of competition from species with narrow ranges of foods, ipso facto necessitates his having access to a matching broad spectrum of environmental resources. Burdened by helpless young and cultural impediments, man is relatively immobile and less able to range widely in search of food and other essentials than are many other mammals. The G/wi are predisposed, therefore, to select as habitats those areas in which the essential resources are most concentrated, that is, the resource nexus. Their restricted choice of residence, combined with the limited number of suitable habitats available in the central Kalahari, imposes on the G/wi the need to regulate the spacing of the population.

The organization of population spacing is based on a number of aspects of sociocultural system. The theological beliefs confer equality of rights to land and its resources on all individuals. Social relationships are largely ordered by the kinship system, which has no lineages and creates a high incidence of status equivalence among individuals. This built-in redundancy facilitates substitution of personnel in groups larger than the household and, combined with the equality of individuals' rights to land and resources, precludes the establishment of exclusive qualifications for membership in the community inhabiting a resource nexus.

The communities are, therefore, open, but the size of each is restricted by the resource state of the nexus at any one time. The G/wi accommodate the annual cycle of biomass fluctuation (and, hence, of the state of food resources) by dispersing the community when the biomass level is low and by local migration in reunited communities when the level is high.

As explained earlier, a band strives to strike an optimal compromise among the four aims of:

1. Obtaining the required amount and variety of food and other resources
2. Exploiting them at the least cost of time and energy
3. So as to retain intact the largest possible residential group that can be sustained by the resources
4. For the longest possible time

These aims are mutually exclusive in that maximizing the attainment of any one lessens the likelihood of attaining the others. To obtain more resources (aim 1) necessitates the expenditure of time and energy (frustrates aim 2). Increasing the size of the residential group (aim 3) shortens the time (frustrates aim 4) before the band members exhaust locally available resources (contra aim 1) and must move camp or cover an excessively large area in their foraging (contra aim 2). The four aims can be represented (Figure 23) as the corners of a square, with the sides and diagonals representing conflict between the aims. The intensity of incompatibility between the aims is pro-

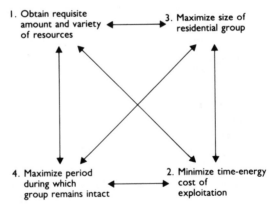

Figure 23. The square of conflicting aims.

A late summer shelter.

portional to the length of the sides. The intensity of conflict will be greatest when resources are scarcest and will be least when resources are most plentiful. For example, it is easier to find food for a larger group and keep that group intact for a longer period when there is a generous supply of food near the camp than is the case in time of scarcity. The availability in the nexus of two (esculent plants, grazing and browsing) of the five categories of resources necessary to survival is directly proportional to the amount of biomass in the habitat (the fourth category, water, is only present in the wet season when biomass is at its maximum). When biomass is least, conflict between the aims, and, hence, the size of the square they form, will be greatest. The amount of biomass and the area of the "square of conflicting aims" are inversely proportional.

By dispersing, reuniting, and migrating, the G/wi keep the "square of conflicting aims" from becoming intolerably large. For instance, when the dry season advances to a stage at which declining esculent plant density forces them to shift the joint camp at inconveniently short intervals, the square has become too large to handle. By splitting into smaller groups, they also reduce the size of the square. The policy of maintaining a favorable energy budget by ensuring a surplus of local plant food over the requirements of the residential group aims, in its effect, at limiting the size of the square.

The household is the most suitable social unit for the period of dispersal. In it there occur in unique combination the relationships of husband and wife, sibling, and parent and child, which make the household sociologically and psychologically the most resilient and durable group in G/wi society and the one in which the short-term needs of its members are most effectively and completely met.

Reduction of the residential group to so small a size (three to six persons) may appear as excessively conservative energy budgeting. By restricting itself to a fixed wintering range, the household is saved the energy that would otherwise be spent on searching for new campsites, moving between them, and building fresh shelters. A unit of three or four able-bodied people (adults and subadult children) is an inefficient scouting force; it takes such a unit a fairly long time to "learn" a new stretch of country, locate the significant growths of food plants, and become familiar with the distribution and patterns of movement of small game for which snares will be set. By retreating into the wintering range before conditions deteriorate to their worst level, the arduous tasks of exploration can be completed while food is still relatively plentiful. As the food becomes scarcer and harder to find, the scouting force learns more and becomes progressively more efficient in its search for food. By September or October it becomes clear that the conservatism of the energy budget has not been at all misjudged. An obvious loss of weight and the appearance of what seem to be the prodromal signs of heatstroke suggest that the "square of conflicting aims" that they now face is optimal.

The practice of dispersing the band is functionally analogous with animal hibernation in that the band becomes dormant and inoperative as a social entity. It ceases normal operations when conditions (the onset of seasonal drought) become too stressful for these activities to be continued. Like the hibernating animal that minimizes its respiration losses by reducing its life process to a very low level, the conservative energy budget of the isolated household represents a reduction in the per capita energy cost of maintenance of the band's members. Although the band per se is inoperative, it continues to exist in the minds of its own and other bands' members. It is not dead but retains its identity and status unimpaired and undiminished by its temporary dormancy. Like hibernation, dispersal is a means whereby the G/wi, through drastic energy-conservation measures, are able to extend their habitat and niche space by avoiding some of the intolerable and possibly lethal pressures of the habitat. Hibernation is an escape open only to those animals that have developed the necessary specialized means of making this behavior possible, and the band does have this specialized means. In the same way that a hibernating animal is able to attain dormancy, survive the stress it sought to escape during the period of dormancy, and recover its full faculties when conditions improve, so the band has the capacity to separate and disperse without destroying itself; the households are capable of independent survival during their isolation, and the band can reunite and resume coherent, coordinated social functioning when the time is appropriate.

Another institution that allows the G/wi to extend their habitat is the system of alliances between bands. Despite the formal analogy that I have drawn between social hibernation and animal behavior, both social hibernation and the habitat-extension function of band alliances are uniquely human in that the organizational framework of the dispersed band and the links constituting the alliances exist only in the minds of the participants and are symbolically maintained. Alliances provide the band with an escape from localized drought in the home territory when more favorable conditions prevail in an ally's territory. The arrangement also provides a means of recovery for the survivors of catastrophic mortality caused by widespread droughts or epidemics. Remnants of decimated bands use their alliances as vectors of regrouping to restore numbers to a level above the threshold of viability. The prior existence of relationships between respective members not only facilitates the integration of the survivors into a new community but also lends them the comfort and strength to recover from the horror through which they have passed.

The network of direct and indirect contacts that alliances provide stretches across the central Kalahari to eventually link, albeit somewhat tenuously, the most remote bands with one another and with the rest of the country. Trade goods and information move across the network and become accessible to the G/wi without requiring them to do more than a fraction of the traveling that would be necessary if they had actually to go and fetch and hear things for themselves. Although the flow of information in and out of the central Kalahari is negligible (e.g., nobody could tell me where Bushmen were to be found in that region and few people even knew that any were there when I started my study in 1958), it is considerable within G/wi country and is an important input to their understanding of the current state of their physical and social environment. (An aspect of the role of information in interband relationships is discussed in greater detail later, under "Opportunism, Cohesion, and Information."

Alliances provide the main means of overcoming demographic imbalance within the band by extending the deme to include the membership of other bands. Even in a group of allied bands, the sex ratio of children was such that several would have not been able to find spouses of suitable age if marriage were confined to that alliance (see the section "Kinship and Marriage" in Chapter 4). Within the confines of the band the probability of unfavorable sex ratios and age combinations is, of course, considerably higher. The demographic dead end is avoided by the fact that alliances are not closed groups (in which each band in a set would have the same allies as any other band in the set) but are open (in which no two bands in a set have the same allies). Furthermore, alliances are potentially universalistic from the individual point of view. In the same way that the kin of my relative can be deemed by kinsmen, I have an entrée into the bands that are allied to my ally.

Operation of any one of the functions of an alliance reinforces the potential for operating any of the others. Intermarriage encourages the exchange of visits and, consequently, of information and goods. Giving succor to an ally afflicted by local drought increases the opportunity for exchange in the future and so on. Alliances are a highly efficient adaptation to the need to space population in times of unusual environmental stress. Alliances provide an extremely economical means by which band members have access to material, informational, and social resources far beyond their territory's boundaries; at the same time, however, they are not required to contribute any more to the maintenance of those resources than they devote to their own band's resources. It is not necessary for them to occupy the areas containing the distant resources because this is done on their behalf, as it were, by the allies (and the allies' allies). Lee (1976:74–97), in discussing interband relationships among the !Kung, has argued that in the long run of 50 to 200 years a group probably requires access to an area ten times the size of that which it covers in a single good year if it is to survive the extreme environmental fluctuation characteristic of the Kalahari. Relationships between !Kung groups have much in common with G/wi alliances and it is clear that both Bushman peoples have devised much the same social solution to this problem. I would emphasize, however, that alliance gives access not only to additional sources of food but also to social and informational resources, which, in their way, are useful to survival as are material resources.

Mobility

The success of the population spacing mechanisms depends on the mobility of individuals and the community. Resource states of all territories vary from year to year with the rainfall experienced in each, and population spacing needs to be sufficiently labile to adjust to these variations. Their technology does not enable the G/wi to manipulate resource states directly; they can do so only indirectly by adjusting the population load that the resources have to bear by transferring population to an area where the best supplies can most conveniently be obtained, whether within the home territory or in another. As I have explained earlier, other economic and social reasons (exchange of goods, services, information; the search for marriage partners; maintaining social contact etc.) facilitate mobility and interband movement as well.

Physical mobility is enhanced by the small inventory of artifacts with which the G/wi make do and the consequently manageable load that a household carries when on the move. By restricting travel to the times when a good en route food supply is available, they obviate the need for carrying rations with them and thus further enhance mobility.

Mobility between one social milieu and another is a function of the individual's versatility, the ability of his previous group to adapt to his absence, and the ability of his new group to adjust to his presence. The use of the kinship system as an organizing principle reduces the extent of necessary individual versatility by providing a set of values and behavioral patterns that are relatively standard for all bands. The wide scope of the system assists the adjustment of both newcomer and the group by furnishing him with an almost ready-made place as a kinsman (real or fictive) of at least one member of his new group. Band adaptation to the arrival or departure of members is facilitated by the "mechanical" nature (see Durkheim, 1964:70–110) of G/wi society in which there is little formal differentiation of statuses. Departure does not, therefore, leave an organizational gap, and the arrival of a new member does not require any extensive rearrangement of statuses and roles to make a place for the newcomer.

A household's capacity for short-term (i.e., up to 8 months) economic autonomy contributes to its social mobility by lessening its dependence on other households. This capacity is derived from the lack of specialization and the fact that the combined manpower of the household possesses all esssential skills. Psychosocial barriers to mobility exist and are manifested in the reluctance of households to sep-

Isolated two-shelter household during separation phase of yearly cycle. The Bushman Survey vehicle is in the background at the right.

arate in winter, as well as in the protracted process of the hiving off of a daughter band from the parent band. The inertia that these barriers induce is overcome by economic pressures and by the thought that parting is not permanent; the sadness of departure is alleviated by anticipation of reunion in the next summer or during the visiting season.

Hygiene

Maintenance of G/wi hygiene depends largely on the mobility and spacing of the population. Household detritus accumulates rapidly as inedible portions of animals and plants are discarded. Rubbish is simply pushed to the edge of the courtyard to be mixed with ash swept out of the fireplace. Band members defecate on the open ground about the periphery of the camp. Swarms of flies and other insects are attracted by the rubbish and feces and constitute an effective series of vectors of infection. The danger is reduced in summer when dung beetles (various Scarabaeidae) are active, as they soon remove or bury fecal material. Dust from ash heaps flies about with

284

every puff of wind to irritate eyes, throats, and lungs and aggravate any ophthalmic condition or respiratory infection. Huts provide shelter and breeding sites for external parasites and soon become infested. Campsites are usually noticeably polluted within 10 days of occupation and after 3 weeks pollution reaches the level of discomfort (judged by informants' standards).

The G/wi like to sit close together and interpersonal contact is frequent and extensive. As a vector of infection, such contact is reinforced by the large number of articles that are in circulation in the band, thus maintaining an effective network of interpersonal vectors. This network is extended to new campsites by the movement of those already infected and of the articles they carry with them but is attenuated by the absence of pollution, which also reduces proneness to fresh infection.

Although the spacing of the population in discrete band territories undoubtedly inhibits the spread of some diseases between communities, it is not an entirely effective barrier, as is evident from the proportions of the 1950 smallpox epidemic. Spacing does tend, however, to localize outbreaks of influenza, the common cold, anthrax, and other diseases of relatively short course.

Demographic aspects

The G/wi preference for first marriages between "cross-cousins," that is, where the bride and groom and their respective parents are all well known to one another, tends to localize the field of spouse selection to the home and allied bands. In this respect the G/wi system contrasts with, for instance, some of the Central Australian Aboriginal marriage systems (see Yengoyan, 1968; Silberbauer, 1971:29–30), which exert centrifugal pressures. The endogamous tendency of the G/wi system functions to reinforce intra- and inter-band bonds and to intensify existing relationships, rather than as a mechanism for establishing and maintaining relationships between more widely separated communities through exogamy. The G/wi system is, nevertheless, very open and is not necessarily restrictive in its operation. If no suitable spouse is locally available, the elastic web of kinship and the widespread exchange network provide avenues for widening the search to cover, at least potentially, the whole G/wi population of appropriate sex and age and even to extend to neighboring non-G/wi Bushmen.

Hypothetically, the localizing function of the G/wi system and its tendency to restrict the size of the deme could lead to a measure of inbreeding if the population were geographically stable. The number of "cross-cousins" who, in fact, had common grandparents would tend to increase. However, there are several factors that counter this tendency. The overall sex ratio is about 93 (Campbell, 1965:29; Silberbauer, 1965:16). If a ratio of 100 obtained in each band, the kinship system is such that the population would be almost evenly divided into preferred (43 percent) and prohibited (57 percent) spouses, and the localizing tendency of the marriage system might produce inbreeding. However, in a sample of 104 households in bands that were linked by alliances, the sex ratios of children categorized by age was:

> 0 to 5 years: 135
> 6 to 10 years: 187
> 11 to 15 years: 50

This local imbalance, in the light of the prevalence of monogamy, creates its own tendency toward exogamy.

The sample is unrealistic in that it does not take into account the full range of alliances of each band; but it does serve at least to indicate that in one linked population the G/wi marriage system would be unable to operate in a localized deme of these dimensions. The sample also takes no account of the "mixing" effects of normal interband movements and catastrophic episodes (e.g., the 1950 smallpox epidemic), which have periodically decimated and redistributed populations.

The model by which Neel and Salzano (Salzano, 1972:235) describe the pattern followed by the genetic structure of hunter-gatherer populations does not fit the G/wi case, although the overall rate of genetic admixture probably approximates that predicted by their model. Inbreeding does occur, but its incidence is not sufficient to produce any outwardly apparent local reduction of variation from the mean phenotypic value.

Early marriage and permitted polygamy maximize the probability of a woman's being married during the whole of her reproductive life. No instance was discovered of a woman's remaining single during that part of her life. In social terms, breeding capacity is maximized. However, natality is not particularly high. In a 1964 sample of 52 women who had passed reproductive age (judged by their own statements and the fact that none had young children) the average number of live births per mother was 2.94. The average number of children reared to puberty and marriageable age was 2.77 per mother. Modal natality was 3 per mother. The rather low child mor-

Table 10. *Ages of 167 persons in three bands and of 1151 Bushmen in Ghanzi District*

Age (yr)	Percent of persons of each age in	
	Three bands[a]	Ghanzi district[b]
0–1	7.8	
2–5	20.0	19.5
6–10	9.8	14.7
11–15	7.2	7.4
16–20	11.2	12.0
21+	44.0	46.4
Total	100.0	100.0

[a] Data from author's survey, 1961.
[b] Data from national census, 1964.

tality (7 percent) indicated by the mothers' reports was confirmed by reference to other informants, by comparison with the histories of younger mothers, and by observations made between 1958 and 1966.

In 1961 I recorded the age structure of the populations of the ≠xade, Easter Pan, and G!õsa bands (then totaling 167 persons). For comparison I took a sample of 1151 from the Bushman data for the whole of Ghanzi district (which includes the central Kalahari) from the national census of 1964. The data are presented in Table 10.

The small percentage of young adolescents in both samples and the high proportion of younger children in the 1961 sample are probably aftereffects of the 1950 epidemic. Mortality appears to have been high in all age groups, the lowest occurring among those then aged 5 to 15 years; to judge from the large number of pockmarked individuals in that section of the population, it seems that many contracted the disease but subsequently recovered. Ten years later, in 1961, this group had reached reproductive age. Differential mortality appears to have produced a situation in which an unusually large proportion of the population consists of young, breeding adults (22.2 percent of the sample), accounting for the "baby boom."

The smaller proportion of children aged 0 to 5 and the larger proportion of young adolescents in the 1964 sample are accounted for by the interval of 3 years separating the two samples, the rather higher marrying-age of non-G/wi, and the fact that the epidemic did not spread to all Bushmen in the district and left untouched many included in the 1964 sample.

Individuals older than 21 are grouped together for purposes of comparison of the samples. Most of the enumerators in the 1964 census came from non-Bushman parts of Botswana and were not experienced in judging the ages of Bushmen. Their estimates of the ages of older adults were too unreliable for use. The census categories did not distinguish those in their first year of life from the group aged up to 5 years, and both are, therefore, listed as one category in the second, larger sample.

I experienced difficulty in judging age myself. There are no event calendars to fix dates, and although it was possible to estimate the ages of children and younger adults with fair accuracy by a lengthy process of analyzing relative ages and working back to the recent, identifiable past, this procedure became less reliable with the increasing age of subjects. Life expectancy in the central Kalahari is, therefore, difficult to calculate. The indications were that few people reach an age of much more than 45. This contrasts with Lee's (1972) figure of 9.9 percent (46 individuals) aged 60 years and over among the !Kung. Lee's data are likely to be correct, as the !Kung with their longer and much more intensive contact with non-Bushmen do have event calendars. For instance, Lee told me of a !Kung woman who was a young girl at the time Hendrik van Zyl (see Chapter 1) hunted elephants in the /ai/ai region of Ngamiland, which would have made her perhaps 100 years old in 1964. If my lower figure is valid, the difference might be explained by the greater environmental stress that the G/wi experience in early summer or, perhaps, by the effects of the smallpox epidemic.

Mortality was very low between 1958 and 1966, and I have little direct information about causal factors operating over longer periods. In my experience one man and his child were killed by a lion and one child died of burns. Two children died of what was tentatively diagnosed as smallpox on the basis of my report, and two adults died of anthrax. In the vernacular image, drought and subsequent famine are major causes of death. I could obtain no quantitative data and misgive the actual statistical validity of the weighting that G/wi cognition accords this factor in mortality. Undoubtedly Bushmen do succumb in years of very serious drought; in the 1939 drought J. D. A. Germond, formerly divisional commissioner, Southern Protectorate, encountered a seriously dehydrated party of 47 without any food or water. Despite his efforts and those of his companions, 37 of the Bushmen died within 24 hours (J. D. A. Germond, pers. comm.; see also Bleek, 1928:40).

During the Bushman Survey, medical treatment, including hospitalization, was made available where necessary. In a number of instances the condition of patients was so serious that it seemed certain that their lives had been saved by the treatment. However, this is drawn into doubt by comparison of places and times when medical treatment could not be given; there are no significant differences in actual mortality. The powers of recovery of G/wi patients, assisted by G/wi medicine, are apparently greater than had been assumed.

Catastrophic reduction and disruption of the population, such as occurred during the 1950 epidemic, are evidently not unique. Mortality during the 1918 influenza epidemic was reportedly high (Dornan, 1925:141). Smallpox swept through southern Africa in several waves in the past (Mackenzie, 1871:250 ff.), and the 1858–63 epidemic wrought havoc among Bushmen in the Kalahari (Tabler, 1955:147). There were also several measles epidemics (Mackenzie, 1871:250 ff.). The rinderpest epizootic at the turn of the century (Henning, 1956:828–841) drastically reduced the ungulate and other game populations of the Kalahari and brought famine to the Bushmen (Passarge, 1907:7 ff.). More serious famine accompanied the droughts of the past (Kokot, 1948).

Passarge's (1907) mention of the G/wi is the only direct evidence of the length of time they have spent in the central Kalahari. They have no memory in the form of oral history or legend of their entry into this area but "have always been there." Their language is dialectally distinct from those of their neighbors and has features intermediate between them. The uniformity of the dialect "gradient" indicates a long correspondence between linguistic and geographical location. There is no evidence of significant immigration of other peoples in the G/wi communities. The conclusion drawn from these facts is that the present population has been established in the central Kalahari for a long time and that it has recovered from the periodical catastrophic reduction of the past at least mainly, if not entirely, through its own ability to survive and regenerate. From the age structure of the population it appears that recovery, although initially slow, commences soon or immediately after abatement of the catastrophic factor. There is not the long-term decline characteristic of a population that has outgrown its resources (see Stott, 1962:355–376; Turnbull, 1972:310–311).

The long-term population curve of the G/wi has, it appears, the shape of a ripsaw and is comparable with the curve computed by Langer (1971:35) for the eleventh- to eighteenth-century population of Europe, except that the G/wi curve has a horizontal, not a rising, overall trend. This is congruent with Lee and Devore's (1968:11) suggestion of periodical ecological reverses that cut back hunter-gatherer populations and Birdsell's (1953:84) dictum "that these populations must have been in essential equilibrium with their environment." The picture is of a population growing at an increasing rate but leveling off before the population outgrows its resources (i.e., before population density exceeds long-term subsistence density, to use Dasmann's term [1964a:183]) as the proportion of breeding adults declines. When, as in the case of drought, it does exceed its immediate subsistence density, it is drastically reduced by epidemic or drought-induced famine. Colinvaux (1973:358 ff.) has argued at length that catastrophes that produce sawtooth population curves are not really density-dependent phenomena because their processes continue to depress the population to levels lower than those at which the catastrophe occurred. I take this point and add that density dependence, in the strict sense, requires that, other factors held constant, density be both a necessary and sufficient cause of the phenomenon. If this were so, the drop in the sawtooth curve would extend no farther than the level of critical density and the relationship between phenomenon and density would be parallel to that between the melting temperature of a substance and its change from a solid to a liquid state.

I suggest instead that the processes under examination here, epidemic and drought-induced famine, which bring about catastrophic declines represented by ripsaw-tooth curves, are density triggered. That is to say, they are initiated when population density exceeds a certain threshold, but once started, they no longer require density as a condition of their operation. The threshold is not fixed but is a function of ambient circumstances; density of x per km^2 may be safe in one year and disastrous in another.

It is easy to see how an epidemic continues to take its toll even after population density declines. Sufferers contract the infection from earlier victims who die during the incubation period of the disease in the former. Yet other victims die of complications caused by the debilitating effects of their illness, and some are victims of neglect, including starvation, because the survivors cannot manage the extra burden of giving them the care needed for their recovery. Consequently, the epidemic continues to take its direct and indirect toll long after population density has declined to a level at which no epidemic would have occurred. Mortality in drought-induced famine persists in similar fashion. These are, if you will, chain reactions triggered when a level of population density is exceeded.

We have a history, then, of a population that grows, is drastically reduced by drought or epidemic, recovers to recommence the cycle, and, over long periods of time, appears to have maintained a fairly uniform series of these violent fluctuations.

Opportunism, cohesion, and information

The G/wi do not confront their habitat in an attempt to manipulate or modify it to any extent in wresting a living in and from it. Instead, they are opportunists who watch and, having read the signs, dispose themselves to use the fluctuations in their habitat to their own best advantage. They strive to maximize their gains not by inducing change but by exploiting the most productive resources when they are at their peak, and when these have passed their zenith, they shift their attention to what next becomes most abundant. In a sense, their way of taking what the habitat most easily yields makes them specialist opportunists.

Opportunism lies in keeping open the widest possible range of options, pursuing none to the point of closing off too many others and retaining a capacity to switch to some other course of action. To manage this strategy requires a close correspondence between actions of the band and developments in the habitat. The presence of a fair number of esculent plants, whose respective periods of availability together cover all seasons of the year, and a variety of herbivores with high meat yields does not mean that the G/wi can passively wait for the habitat to provide these things. They must plan their moves and be in position to take the advantage when opportunity presents itself. This, of course, is true of most, if not all, hunter-gatherers. Although G/wi opportunism in the totality of its features is unique, other hunters and gatherers have their own opportunistic modes.

The G/wi view of the world, in which man is the inventor and discoverer who must devise his own modus vivendi within the constraints of coexistence with N!adima's other creatures, provides a framework for action. Within this framework, the G/wi are allowed to hunt and gather as much as they need but no more and never to an extent that might bring about the extinction of other species. This gives them a catholic choice, which is restricted only by a few taboos (these are age specific in relation to people and cover snakes and some antelope) and their own capabilities, preferences, and prejudices.

A formidable extent and level of capability are necessary to an optimal mode of crop switching. It requires a sound knowledge of field botany so that one can not only recognize the plants but also avoid a ruinous waste of time and energy in searching for them, that is, their distribution within the band territory must be intimately known. A thorough understanding of the effects of climate on the size and quality of each relevant crop in each part of the territory is the basis of planning an effective program of migration and dispersal of the band. The bow and arrow are versatile weapons, but switching from one prey species to another involves more than simply aiming at a different animal. Each species calls for a different technique of approach, of target selection, and, because of the differences in susceptibility to arrow poison, a different method of pursuit after the arrow has struck home. All these factors must be assessed when choosing among the available options and must be included in the calculation of how and when to exercise a particular option without excluding too many others.

In assessing the opportunity cost of an option, the G/wi are inescapably bound by the relationship between commitment to a line of action and the rewards to be gained from it. As opportunists striving to maximize choice, they cannot commit themselves too far in any direction and must therefore accept commensurately low returns. In other words, as generalists they must forgo the benefits of specialization if they will not pay its prices. Specialization is possible in this habitat, as is demonstrated by the pastoralists and ranchers in other, comparable parts of the Kalahari. But it is not feasible with a hunter-gatherer's technology. With that technology, the nonspecialist trades off the benefits of specialization against the small returns of the opportunist who will not commit himself. These small returns are tolerable because, in most seasons of most years, they come in a fluctuating but adequate flow of food and other goods that meets the requirements of his life-style. For hunter-gatherers, whose survival depends on the health of their habitat, an opportunist strategy of crop switching and prey species alternation minimizes the impact of their gathering and hunting on the integrity of the habitat.

Food preferences and prejudices lead to an eclecticism that slightly deflects the opportunistic aim of taking the most from that which is currently most abundant. I have explained the G/wi scale of preference among food plants; they rate highest the plants that yield the greatest amount of food for the least effort expended in gathering. This is, of course, consistent with an opportunist strategy but does not always mean that the most abundant species is the most heavily exploited. The conservation measures of saving *Raphionacme* and *Coccinia* tubers for the dry season is a further exception. Although popular literature and non-Bushman folklore present the Bushmen as omnivores, they are as fastidious in their choices as any other people. Carnivores in general are seldom hunted, although the meat is enjoyed when it becomes available (e.g., man-eating lions and leopards that I shot), and other Bushmen (and G/wi visiting these people) regularly eat the meat of foxes, which are hunted for their pelts. Small birds and rodents are hunted by children but are not considered fit food for adults (which was also the attitude of our parents when, as small boys, we investigated the culinary qualities of such things as mice, sparrows, and locusts). They also show a conservative reluctance to hunt animals that are only sporadically encountered. It is admirable common sense not to shoot a flimsy arrow into dangerous quarry like elephant and buffalo unless you know exactly what the prey will do next (both species take their annual toll of hunters armed with high-powered rifles and give rise to innumerable tales of narrow escapes), but why zebra were not hunted I could not determine.

Although opportunism is the antithesis of specialization, it paradoxically requires its own type of specialized activities and sociocultural accommodation. To take advantage of environmental fluctuation and manage the "square of conflicting aims," the G/wi have developed forms of social specialization that make migration, dispersal, and social hibernation possible. The facility with which separation and reuniting are performed derives from several features of their social organization. The common set of definitions and models of status and behavior that the kinship system provides constitutes a broad framework within which an individual can choose his course of action and predict and interpret the behavior of others and the group can evaluate and comprehend the actions of the individual. As a code of behavior, a kinship system has two characteristics that are particularly apt. First, it imposes lifelong relationships, which, although they may be inactive through lack of access, cannot be denied or erased once interaction does occur. Second, the way in which kin categories are defined constitutes a cognitive map of one's own position vis-à-vis others and also of the juxtaposition of other to other.

As is commonly the case, the household and its internal relationships are accorded unique character. They can be transferred or conferred only by the special events of marriage (and death or divorce). In the processes of fusion and fission of the band, the household is functionally an irreducible unit. By contrast, there is a high incidence of equivalence among extrahousehold relationships and a great measure of versatility among the corresponding roles. Bonds of friendship and special affection may reduce the degree of real substitutability of some individual extrahousehold relationships, but the level of redundancy among them is generally such that, as I have explained earlier, one or a number may join or leave a band without disrupting the structural order. New combinations of households are thus easily formed and existing combinations are easily dismantled. To borrow the chemist's term, groups of households have high valency in processes of fusion and low inertia in processes of fission. Fission inertia is further suppressed by the absence of a centralized, formal authority structure in the band, dependence on which would otherwise inhibit dispersal.

Human behavior is largely learned and must, therefore, be directed by rules (and values for interpreting the rules). These rules must be learned and shared by all members of the group if their actions are to be coordinated and comprehended instead of random and mystifying. Social anthropology has developed from the study of societies that are small-scale and simple, in comparison with metropolis-oriented societies, but even these are of much greater scale and complexity than are hunter-gatherer bands. The small states, tribes, and peasant villages to which social anthropology has given much of its attention have social systems in which structural constraint of behavior is more directive than is the formal organization of a band. The lability, versatility, and redundancy that are characteristic of G/wi interhousehold relationships preclude structural constraint of this extent, and the regulation of behavior in the band must be explained in terms that go beyond the traditional areas of focus in social anthropology. The facility of fusion and fission is gained at the cost of an adequate formal social definition. In the intimate life of the band, relationships develop a rich and diverse quality of expression that requires an elaboration of the social fabric beyond the simple pattern prescribed by the kinship system. When households are camped together, the band members need a means to define more precisely the current gradations in social and emotional closeness and distance that distinguish bonds of greater and lesser intensity. Such means must be responsive to the dynamic nature of the relationship web, be able to accommodate households joining and leaving, and be able to express the rise and fall of emotional level and intensity and the variations in the quality of interaction.

Social definitions of status, closeness and distance, and intensity of bonds are parcels of information conveyed in the learning of roles and in the interaction between those to whom the definitions refer (and, fortunately for the anthropologist, are also conveyed by observing such interaction). In this context the sense in which the term "information" is used is derived from the cybernetic meaning of the term, namely, a logarithmic measure of the improbabilty in a given situation of the occurrence of an event. To state that the improbability of an event approaches zero is to assign high probability to its occurrence. If the truth of one among a set of mutually exclusive propositions depends upon the occurrence (or non-occurrence) of an event, then information is that which enables one to select the correct proposition by assigning a measure or probability to the occurrence (or non-occurrence) of the event. In short, information is that which reduces uncertainty.

Information is required to elaborate the social fabric so as to express nuances of relationship. A structure that depends on redundancy to facilitate substitution is obviously lacking in intrinsic information for this purpose; to distinguish among relationships that are essentially similar in form, additional inputs of information are required. In the band the source of this information is the exchange of goods and services between households. Exchanges of prestations both define and express relationships. As I have explained, the capacity of the G/wi exchange system to equate a broad spectrum of goods and services for purposes of reciprocation and the relativism in its scale of values facilitate the flow of commodities and favors. The rate of flow between two participants in the system is not appreciably determined by, nor does it have a significant effect on, the rate of transactions between other participants. People, therefore, have a good deal of freedom to distinguish among the other households in the quality and quantity of their exchanges and to thus define and express the relationship. This same freedom allows the system to expand and contract, without loss of coherence, its capacity to handle transactions (within limits, as discussed later) according to the number of people participating in the exchanges. These features make the exchange system an effective source of information and channel for communicating those facts needed to compensate for structural redundancy.

To act as a cohesive, coordinated entity, the band requires its own level of ordering. Successful execution of a strategy of opportunism demands flexibility of organization, versatility, and a readiness to make rapid changes of direction and, hence, frequent decisions and a high level of solidarity in effecting them. Band consensus, as a style of band politics, can reach decisions with the requisite rapidity and freely allows changes of earlier decisions when subsequent events or later intelligence show them to be inappropriate. This means of reaching decisions places responsibility for community government on each member, ensuring support for the course of action decided upon, however frequent the changes may be. The consensus polity, like the exchange system, can match its capacity to the number of participants, for each household comes, as it were, complete with all the necessary political apparatus. Organizational capacity increases with community size (again, within limits) but without a concomitant need for superstructure additional to that provided by the households joining the community.

This is a neat way of conforming to Ashby's law of requisite variety, which states that the variety of alternative control actions that a regulating device is capable of must be at least equal to the variety of fluctuations and deviations produced by that which is controlled – a regulating system must be able to generate as many states as can the system that it controls (Ashby, 1956:202–218). The organizational capacity generated by the household-in-community is at least equal to the increment in variety in behavior occasioned by its joining the community. It will be appreciated that the inclusion of additional households gives scope for behavior that would not be possible in a smaller social circle. Complexity and variety increase most rapidly in the approximate range of three to six households and then more slowly as the relative significance of each further addition to the community diminishes.

Organizational capacity generated by the household-in-community is manifested not only in political processes but also in such aspects as the activities of joking relatives, in exorcising dances, and in the operation of other means of social control. In this context joking relatives function as feedback loops that provide an individual with information about others' reactions to his behavior, reducing randomness both in his behavior and his anticipation of its social consequences. The avoidance/respect relationship is narrower in scope, having smaller propensity for negative feedback than there is between joking partners, but, particularly in regard to prestigious avoidance/respect relatives, there is a good deal of positive feedback. Both the exorcising dance and joking relationships have a rather unusual information function in erasing, as well as countering, entropy in the social system. What I mean by erasure is that both serve to diminish resentment (which I see as entropic) of others' offenses by redefining the situation or by directing resentment to harmless dissipation outside the social system. The exchanges of joking partners include opportunity for making amends and for extending forgiveness, which, because it is a public performance, gives emotional comfort not only to the partners but to the whole band as well. Successful exorcising dances remove the bitterness of others' misdeeds by expelling "G//amama's evil."

Whereas processes of fusion are accompanied by increased demand for, and provision of, information, processes of fission are accompanied by (and, I think, triggered by) a reduction of information and an increasing entropy in the community or group concerned. The end of a band's visit to an ally, described in Chapter 4, is an account of mounting tension and incipient conflict that is symptomatic of an ebb in the flow of information in the system (here, a two-band encampment). The ebb is occasioned by the nature of the information initially generated by the visit – pleasure at reunion, news of happenings since the last meeting, exchanges prompted by the visitors' gifts, all have a limited life. Away from their home territory the visitors cannot continue the series of gift exchanges. Lacking the shared, permanent common purpose characteristic of band life, they cannot integrate sufficiently with the host organization to render a significant number of acceptable services, participate meaningfully in political processes, or engage more than superficially in most of the day-to-day social interaction among the host households. Having the status of visitors inhibits behavior to some extent and precludes or stultifies many of the normal maintenance activities. The presence of visitors is comparably disruptive of processes in the host band. Information generated by the arrival of the visitors is insufficient to arrest entropy into the increasingly irregular and erratic social, political, and economic environment of dual band coexistence. Finally, there may be the entropic tendency of the threat of coming food shortage if the visit is unduly prolonged.

The looming threat of food shortage is, of course, what triggers fission in the united band and sends the households off to their individual winter ranges. It is not of much use beyond the context of this discussion to do so, but food shortage can be viewed as a powerful information-destroying entropic force. It is the factor that arrests fusion processes at the level of the clique in those drought years when the band is unable to unite and is compelled to remain grouped in coteries occupying separate camps.

Processes of fission in the household (excluding death, which is an event rather than a process) are also initiated by information deficit. Divorce is preceded by increasing randomness in the behavior of a husband or wife, or both, and disturbance of the normal information content of roles by denying expectation of role-appropriate behavior. Eventually entropy in this form reaches an intolerable level and one of the spouses leaves. Efforts by band fellows to rescue the marriage consist of attempts to guide the spouses' behavior by persuasion and threat, emotional support or deprivation, and interpretation of a spouse's behavior that presents it in a favorable light for the other. These all constitute inputs of information to reduce perceived randomness of behavior. If adequate, the measures reduce entropy to below the threshold of fission and the marriage survives.

It would be tedious to analyze states of information networks during the phases of the developmental cycle of the household, in clique disassociation, and in the processes by which a band "eases out" an intractable, unwanted member. My point is that the balance in supply and demand of information is a necessary but not sufficient factor (hence, a limiting factor) in the fusion and fission processes on which the G/wi strategy of opportunism depends. This has relevance to the socioecology of other hunter-gatherers. Turnbull (1968) has discussed the importance of flux (i.e., fission and fusion) in Mbuti and Ik groups. From this it appears that fission in these societies is also triggered by rising levels of entropy or by diminution in the flow of information in the sociocultural system. In contrast to the G/wi, Mbuti net-hunters disperse during what they regard as a season of plenty, the honey season. As Turnbull (1968:135) explains, they do this because the bountiful food supply removes the necessity for tight cooperation, which exists for the rest of the year. This tight interdependence is analogous with the high intrinsic information content of the complementary roles of husband and wife and their interdependence in the isolated G/wi household. In the Mbuti case the inputs of compulsory cooperation cease to operate when the honey season commences, and rising entropy leads to fission. The Mbuti archers follow an exactly opposite program, banding together during the honey season and for precisely similar reasons; *they* see this as a time of food shortage that demands maximum cooperation (Turnbull, 1968:135).

It appears from the cases of the Mbuti, Ik, and G/wi that there is an underlying tendency to fission and anarchistic dispersal. This is a matter of perspective; fission is a reflection of the tendency present in all systems to develop entropy. It seems more fruitful to argue that there is a tendency to arrest entropy in sociocultural systems and to maximize group size to the extent that the information capacity of the system permits. This being so, the G/wi in particular and band societies in general are limited not only by environmental factors in the size of the residential group that can be sustained under a given set of conditions but also by the structure, topology, and state of the information network of their sociocultural system. Widely divergent views have been expressed on the matter of the carrying capacity of the habitats of hunter-gatherers and the rate of exploitation for stability of these populations. Lee and Devore (1968:11) have suggested that successful long-term exploitation would be at a rate of 20 to 30 percent of the carrying capacity; Birdsell (pers. comm., 1973) believes the rate is on the order of 95 percent. This debate does not seem to me capable of resolution in this form. Apart from the difficulty of establishing what the carrying capacity of a given habitat may be (it is a long-term measure, unlike subsistence density, which is an ephemeral rate), the arid environments inhabited by a large proportion of the hunter-gatherers for whom we have good data are subject to the same sort of seasonal and long-term variation as I have described for the central Kalahari; in addition, we have little understanding of the consequences of extreme drought for hunters and gatherers. I do not believe we are in any position to accurately assess carrying capacity. We might find it more rewarding to look at information networks as limiting factors in community size because it seems to me unlikely that the folk wisdom of the G/wi, for instance, would retain a memory over successive generations of the land's carrying capacity in times of record drought and use this as the datum for setting the upper limit of their group size, population density, and exploitation density.

It is clear that G/wi social organization can order a community of 85 members (the ≠xade band) indefinitely. A camp of 138 allies remained ordered for only a few weeks. The greater number gathered around the borehole in 1962 was chaotic. Perhaps the upper limit of a permanent community is something like 90 to 100. I have said that the balance between the supply and demand of information is a factor limiting community size but that social structure is not determinate in this way. It seems possible that the limits to which households-in-community can generate information and organizational capacity are set by the topology of relationships (as distinct from the structure of relationships; topology is concerned with the properties of figures that persist despite deformation – structural properties are altered by deformation).

Most, if not all, significant interaction in a band is either direct, that is, face-to-face between two or more persons (call this first-order relationship), or indirect, that is, through one intermediary who is well known to the two actors (second-order relationship). In a band one either tells or asks somebody something oneself or one asks somebody to ask another person. One does not ask somebody to ask another person to tell a third person something (a third-order relationship or interaction). As a channel of communication, a control mechanism, and a means of error correction, this is very efficient. First-order interaction between people who know each other as intimately as do band fellows contains a flood of information. (To illustrate the point: When my wife asks me where I put the what'sitsname with the thingamabob on it or suddenly resumes, in mid-sentence, a conversation interrupted some weeks previously, I quite often know what she is talking about. Only our married friends and those who grew up in large families are undismayed by this sort of knight's-move communication.) Second-order interaction between two band fellows via a third is nearly as effective a channel of communication. (If my wife sends one of the children to ask where I put the whatsitsname, etc., I am sufficiently familiar with the poor child's reaction to having to retain gobbledygook in her memory to know the probable errors in the message that I receive and can correct them. When I send back the reply that it is on top of the whatever that Old Other-thing left behind, my wife can similarly use her knowledge of the child and of me to make sense of what reaches her.) My point is not that the G/wi Bushmen and members of my family communicate in nonsense syllables but that first- and second-order interaction within a circle of intimates is rich enough in information to reduce errors, confusion, and entropy to a tolerable minimum. However, third-order interaction is considerably less efficient as a channel of communication. (If this hypothetical conversation between my wife and me were to be transmitted through one child and then the other, the possibility of error correction shrinks to the point where tempers fray.) The loss in efficiency can be overcome only by restricting the messages to codes that are less versatile and economic but highly specific, allowing less room for error. (My wife is now put to the trouble of not only remembering but also making the child carefully repeat after her the name of the sheep remedy she wants.)

I have used the example of messages passed between my wife and me. Interaction, of course, covers a much wider scope of action and channels of communication, but in all instances more or less information is being conveyed to the recipient with greater or lesser accuracy and precision. The uncertainty in the recipient's mind about the intentions of the actor and the response to his act that is most appropriate is reduced to a greater or lesser extent by the information so transmitted. I have said that most significant interaction in the band is of the first and second orders. It seems likely that there is something in the topology of relationships that limits the number of people who can confine their interaction to these two orders. Only as a metaphor, I refer to the problem of the number of spheres that can be packed around a spherical nucleus so that each is in contact with its neighbors and the nucleus. The answer is given by the formula (radius^2 × 10) + 2. For a layer of 1-sphere depth (or radius), the number is (1^2 × 10) + 2 = 12. This configuration provides the condition of each sphere being in either first- or second-order contact with all other spheres. A 2-sphere layer around the nucleus would accommodate 43 spheres (including the nucleus) and their contacts would be of first to fourth orders. So 13 spheres (12 plus the nucleus) is the maximum number that can be arranged in a contact configuration of first and second orders. Human social relationships are, of course, much more complex than this simple three-dimensional metaphor, and I cannot calculate what the upper limit of a first- and second-order relational configuration would be in a social context. But when it is made, I suspect the calculation would give an answer not far from the 90 to 100 that I suggest as the limit of G/wi capacity.

A round 100 has something of the aura of a magic number in social grouping (e.g., the ninth-century Danish institution, copied elsewhere, of a subdivision of a shire into unites each containing 100 families) and is not, of course, unique to hunter-gatherers, some of whom form social groups of larger than 100. My speculation concerns the manner in which actions of band members are coordinated under conditions of changing size of residential group when there is no accompanying structural change apparent. The upper limit of about 100 may be a chance coincidence with units of that size in other societal contexts, or there may be common factors that explain the frequent occurrence of the number. But that is another matter and, to revert to the G/wi, it is all very well to say that the same structure serves the whole range of sizes of the community: That merely reiterates description. Why is the structure so versatile as to endure the radical deformation imposed by differing group size and why does it come unstuck, like a test cricketer before the wicket, at the century mark? (Although I understand the Witches of Wisden's have shifted the curse to a score of 79.) As previously remarked in this chapter, most of the societies that anthropologists have studied have more elaborate social structures, which more exactly direct social behavior. It is possible that what I have referred to as the topology of relationships is of much lesser importance in the more complex, formally ordered societies.

In discussing political processes in the band (Chapter 4), I referred to Elizabeth Colson's emphasis on dense social networks in small, face-to-face communities as necessary to adequate information flow (Colson, 1974:5, 54 ff.). The density of a substance is its mass per unit volume. In a social network, density refers to the extent to which participants are linked to each other by their relationships; where each has many different types of relationship with all or a large proportion of the others, the network is dense. Clearly this is favorable to the generation and communication of a wide spectrum of information. The correlation of connectedness in a network with facility of communication has been repeatedly demonstrated by social psychologists (see Bales, 1950, 1961; Borgatta and Bales, 1961; Heise and Miller, 1961).

I have said that relationships within the band are built on the model of kinship. This obviously imposes some limitation on the types of relationship that are permitted between various individuals (e.g., joking versus respect), and some relationships are highly restrictive of the nature of interaction (e.g., between one woman and the husband of another). Density is limited, therefore, insofar as variety is restricted, and as channels of communication, some of the more restrictive relationships have rather small capacity. However, where first-order (i.e., direct) contact is attenuated, the structure is such that there is ample "channel capacity" in second-order contact. (Although I must be circumspect in my communicative behavior toward another man's wife, I am allowed considerable freedom when talking to her husband.) The properties of the network, then, are such that those involved can perceive with reasonable commonality the actions that are performed in the band, the circumstances surrounding the actions, and the nuances of meaning that are attributed to both. It follows, of course, that although the use of the kinship system as an organizational framework does impose some restrictions on information flow, it also enhances the informational value of interaction by providing a matrix of meaning for a substantial proportion of transactions in the prescriptions and definitions that are part of kinship. In Ashby's terms, the amount of requisite variety is reduced by the use of the kinship system, correspondingly lessening the rate of flow of information required to coordinate the behavior of the co-resident band members.

I cannot satisfactorily explain the upper limit of about 100 but suggest that this is the largest number of people who can sustain first- and second-order contact in normal social intercourse. In other words, a larger number of people would either have to resort to third order of contact and higher or slow down or otherwise attenuate their interaction with one another. I had thought that the idea could be explored in a fairly straightforward computer simulation, but I conclude that the rates of generation and communication of information in the community must be critical. To perceive and measure these, an observer would need the kind of sensitivity I referred to in the Preface and should also have detailed knowledge of the backgrounds of all members of the band. A simulation would have some heuristic value but, lacking this intimate quantification, would not be appreciably less vague than is the first sentence of this paragraph.

The problem is of some relevance to the broader question of consensus. If it is necessary that all those consenting should have access to a common pool of information – in which must be included the information generated in the process of arriving at consensus – then it seems that consensus, as a mode of decision making, is only practicable in rather small groups and that there might be some critical relationship between the topology of relationships in the group and the requisite rate of communication.

If my speculation has some validity and the topology of relationship is the factor that enables the household-in-community to generate sufficient organizational capacity, then this aspect of G/wi society is an essential element of the strategy of opportunism. Organizational "overheads" are minimized and the range of tactical options open to the band is not restricted in the way that it would be if members had to commit themselves to a hierarchy of formal structures governing their relationships as the residential group changed in size. To illustrate this argument of opportunistic freedom, contrast the band with the organization of a part of an army in the field: Infantry may operate at company strength, detail platoons for action in specific sectors or even detach smaller numbers of men for such tasks as fetching supplies or patroling. The fact that each subdivision of the larger whole is made in accordance with a hierarchical structure of command of diminishing weight and extent means that a lance corporal leading a patrol may initiate only a very restricted range of responses to developments during the engagement. Beyond this range he must await the orders of his sergeant or platoon commander who, in turn, may have to seek authority from *his* superior officer. Keeping the chain of command intact also necessitates that each unit commander give orders only to those immediately beneath him. All this is far too unwieldy for the exercise of opportunistic initiative, which, in any case, is a virtue seldom encouraged in the armed services.

Opportunism in the band, although an effective adaptation to the circumstances of isolated hunter-gatherers, is a highly specialized

303

trait. Depending on a close-knit network of intimate relationships, it makes the G/wi social order singularly vulnerable to dislocation when a more forceful people intrudes. The narrowness of the specialization and its vulnerability are shown by the shocking rapidity and the tragic extent of the collapse of the social organization of the G/wi and other Bushmen when they are overrun by ranchers or cattle-post owners and are reduced to empty, aimless demoralization.

Bibliography

Aberle, D. F., Cohen, A. K., Davis, M. J., Levy, Jr., M. J., and Sutton, F. X. 1949–50. "The Functional Prerequisites of a Society." *Ethics* 60:100–11.

Air Ministry, Meteorological Office. 1960. *Handbook of Aviation Meteorology*. HMSO, London.

Allee, W. C., Emerson A. E., Park, O., Park, T., and Schmidt, K. P. 1949. *Principles of Animal Ecology*. Saunders, Philadelphia.

Anderson, A. A. 1888. *Twenty-Five Years in a Waggon*. Chapman and Hall, London.

Andersson, C. J. 1856. *Lake Ngami or, Explorations and Discoveries during Four Years' Wanderings in the Wilds of South Western Africa*. Hurst and Blackett, London.

Angyal, A. 1969. "A Logic of Systems." In Emery, F. E. (ed.). *Systems Thinking*. Penguin, Ringwood, Australia.

Ashby, W. R. 1956. *Introduction to Cybernetics*. Wiley, New York.

Ashton, E. H. 1937. "Notes on the Political and Judicial Organisation of the Tawana." *Bantu Studies* 11, No. 2:67–83.

Bailey, F. G. 1969. *Stratagems and Spoils*. Blackwell, Oxford.

Baines, T. 1864. *Explorations in South-West Africa*. Longman, Green, Longman, Roberts and Green, London.

Bales, R. F. 1950. *Interaction Process Analysis*. Addison-Wesley, Cambridge, Mass.

Bales, R. F. 1961. "The Equilibrium Problem in Small Groups." In Hare, A. P., Borgatta, E. F., and Bales, R. F. (eds.). *Small Groups*. Knopf, New York.

Barnard, A. J., 1976. "Nharo Bushman Kinship and the Transformations of Khoi Kin Categories." Ph.D. thesis, University of London.

Barnard, A. J. 1978. "The Kin Terminology System of the Nharo Bushmen." *Cahiers d'études africaines* 72(18-4):607–29.

Barrow, J. 1806. *A Voyage to Cochinchina in the Years 1792 and 1793 to which is annexed an account of a journey made in the years 1801 and 1802 to the residence of the Chief of the Booshuana Nation*. Cadell and Davies, London.

Beattie, W. A. 1971. *Beef Cattle Breeding and Management*, 3rd ed. Pastoral Review, Melbourne, Australia.

Benson, M. 1960. *Tshekedi Khama*. Faber and Faber, London.

Berger, P. L., and Luckmann, T. 1967. *The Social Construction of Reality*. Doubleday (Anchor Books), Garden City, N.Y.

Biesele, M. (in press). *!Kung Folklore*. Harvard University Press, Cambridge, Mass.

305

Bibliography

Birdsell, J. 1953. "Some Environmental and Cultural Features Influencing the Structuring of Australian Aboriginal Populations." *American Naturalist* 87(834):171–207.

Blair Rains, A., and Yalala, A. M. 1972. *The Central and Southern State Lands, Botswana.* Land Resources Division of the Foreign and Commonwealth Office Surbiton.

Bleek, D. F. 1928. *The Naron – A Bushman Tribe of the Central Kalahari.* Cambridge University Press, Cambridge.

Bleek, D. F. 1956. *A Bushman Dictionary.* American Oriental Society, New Haven, Conn.

Bleek, W. H. I., and Lloyd, L. 1911. *Specimens of Bushman Folklore.* Allen, London.

Bonsma, J. 1961. Quoted in Palmer, E., and Pitman, N., *Trees of South Africa.* Balkema, Cape Town.

Boocock, C., and van Straten, O. J. 1962. "Notes on the Geology and Hydrogeology of the Central Kalahari Region, Bechuanaland Protectorate." *Transactions of the Geological Society of South Africa.* Vol. 65, pt. 1: 125–71.

Borgatta, E. F., and Bales, R. F. 1961. "Interaction of individuals in reconstituted group." In Hare, P., Borgatta, E. F., and Bales, R. F. (eds.). *Small Groups.* Knopf, New York.

Botswana Archives. File No. S.440/4, 1935. File No. S.360/2, 1937.

Boughey, A. S. 1968. *Ecology of Populations,* 1st ed. Macmillan, New York.

Boughey, A. S. 1971. *Ecology of Populations,* 2nd ed. Macmillan, New York.

Breyer-Brandwijk, M. G. 1937. "A Note on the Bushman Arrow Poison, *Diamphidia Simplex* Peringuey." In Rheinallt-Jones, J. D., and Doke, C. M. (eds.). *Bushmen of the Southern Kalahari.* Witwatersrand University Press, Johannesburg.

Brooks, A. C. 1962. *A Study of the Thomson's Gazelle (Gazella thomsonii Gunther) in Tanganyika.* HMSO, London.

Brown, R. C. 1974. "Climate and Climatic Trends in the Ghanzi District." *Botswana Notes and Records* 6:133–46, Gaborone.

Buchler, I. R., and Selby, H. A. 1968. *Kinship and Social Organization.* Macmillan, New York.

Burchell, W. J. 1822. *Travels in the Interior of Southern Africa.* MacLebose, Glasgow.

Butynski, T. M. 1973. "Life History and Economic Value of the Springhare (*Pedetes capensis forster*) in Botswana." *Botswana Notes and Records* 5: 209–13. Gaborone.

Campbell, A. C. 1965. *Report of the Census of the Bechuanaland Protectorate, 1964.* Bechuanaland Government, Gaberones.

Campbell, A., and Child, G. 1971. "The Impact of Man on the Environment of Botswana." *Botswana Notes and Records* 3:91–110. Gaborone.

Campbell, J. 1815. *Travels in South Africa Undertaken at the Request of the Missionary Society,* 3rd corrected ed. Black, Parry and Hamilton, London.

Campbell, J. 1822. *Travels in South Africa Undertaken at the Request of the Lon-*

Bibliography

don Missionary Society; Being a Narrative of the Second Journey in the Interior of that Country. 2 vols. Westley, London.

Carroll, J. B. (ed.). 1956. *Language, Thought and Reality: Selected Writings of Benjamin Lee Whorf*, Massachusetts Institute of Technology and Wiley, New York.

Chapman, J. 1971. *Travels in the Interior of South Africa 1849–1863*. Balkema, Cape Town.

Clifford, Capt. the Hon. B. E. H. 1928. *Report on a Journey by Motor Transport through the Kalahari Desert, Ghanzi and Ngamiland to the Victoria Falls*. Government Printing and Stationery Office, Pretoria.

Cloudsley-Thompson, J. L., and Chadwick, M. J. 1964. *Life in Deserts*. Foulis, London.

Codd, L. E. W. 1951. *Trees and Shrubs of the Kruger National Park*. Department of Agriculture, Government Printer, Pretoria.

Cohen, G. 1974. "Stone Age Artefacts from Orapa Diamond Mine, Central Botswana." *Botswana Notes and Records* 6:1–4.

Cole, D. T. 1955. *An Introduction to Tswana Grammar*. Longmans, Cape Town.

Cole, M. 1961. *South Africa*. Methuen, London.

Colinvaux, P. A. 1973. *Introduction to Ecology*. Wiley, New York.

Colson, E. 1974. *Tradition and Contract: The Problem of Order*. Heinemann, London.

Cottrell, F. 1955. *Energy and Society*. McGraw-Hill, New York.

Cronwright-Schreiner, S. C. 1925. *The Migratory Springboks of South Africa*. Unwin, London.

Cumming, R. G. 1863. *Five Years' Adventures in the Far Interior of South Africa*. Murray, London.

Dalton, G. 1969. "Theoretical Issues in Economic Anthropology." *Current Anthropology* 10:63–80.

Damas, D. 1969. *Contributions to Anthropology: Ecological Essays*. Bulletin 230, National Museums of Canada, Ottawa.

Dansereau, P. 1957. *Biogeography – an Ecological Perspective*. Ronald Press, New York.

Darling, F. F. 1960. *Wild Life in an African Territory*. Oxford University Press, London.

Darling, F. G., and Milton, J. P. (eds.). 1966. *Future Environments of North America*. Natural History Press, New York.

Dasmann, R. F. 1964a. *Wildlife Biology*. Wiley, New York.

Dasmann, R. F. 1964b. *African Game Ranching*. Pergamon, New York.

Davis, D. H. S. 1946. "A Plague Survey of Ngamiland, Bechuanaland Protectorate, During the Epidemic of 1944–45." *South African Medical Journal* 20:462–67, 511–15.

Davis, D. H. S. 1964. "Ecology of Wild Rodent Plague." In Davis, D. H. S. (ed.). *Ecological Studies in Southern Africa*. Junk, The Hague.

de Klerk, W. A. 1975. *The Puritans in Africa*. Penguin, Ringwood, Australia.

Deutsch, K. W. 1963. *The Nerves of Government*. The Free Press, Glencoe, Ill.

Bibliography

Dice, L. R. 1952. *Natural Communities*. University of Michigan Press, Ann Arbor, Mich.

Donn, W. L. 1951. *Meteorology with Marine Applications*, 2nd ed. McGraw-Hill, New York.

Dornan, S. S. 1925, *Pygmies and Bushmen of the Kalahari*. Seeley, Service, London.

Douglas, M. 1970. *Purity and Danger*. Penguin, Ringwood, Australia.

Duncan, O. D. 1964. "Social Organization and the Eco-system." In Faris, R. E. L. (ed.). *Handbook of Modern Sociology*. Rand McNally, Chicago.

Durkheim, E. 1964. *The Division of Labour in Society*. Collier-Macmillan (Free Press of Glencoe), London.

Easton, D. 1965. *A Framework for Political Analysis*. Prentice-Hall, Englewood Cliffs, N. J.

Eibl-Eibesfeldt, I. 1970. *Ethology: The Biology of Behavior*. Holt, Rinehart and Winston, New York.

Ellerman, I. R., Morrison-Scott, T. C. S., and Hayman, R. W. 1953. *Southern African Mammals*. British Museum, London.

Emery, F. E. (ed.). 1969. *Systems Thinking*. Penguin, Ringwood, Australia.

Energy and power, 1971. *Scientific American* 225, No. 3. Entire issue.

Excell, A. W., and Wild, H. (eds.). 1960. *Flora Zambesiaca*. Crown Agents, London.

Falconer, J. 1971. "Relationships between Wild and Domestic Animals in the Control of Foot and Mouth Disease in Botswana." *Botswana Notes and Records, Special Edition*, No. 1, pp. 153–56, Gaborone.

Feibleman, J., and Friend, J. W. 1969. "The Structure and Function of Organisation." In Emery, F. E. (ed.). *Systems Thinking*. Penguin, Ringwood, Australia.

Fett, M. I. E. 1967. "A Translation and an Appraisal of 'Essay on the Seasonal Variations of Eskimo Societies' (*Année Sociologique*, 1905) by M. Mauss with the Collaboration of H. Beuchat." M. A. thesis, Monash University. Victoria, Australia.

Feuer, L. S. 1953. "Sociological aspects of the relation between language and philosophy," *Philosophy of Science* 20:85–100.

Firth, R. 1950. *Primitive Polynesian Economy*. Humanities, New York.

Fitzsimmonds, V. F. M. 1962. *Snakes of Southern Africa*. Purnell, Cape Town.

Fortes, M. 1966. "Introduction." In Goody, J. (ed.). *The Developmental Cycle in Domestic Groups*. Cambridge University Press, Cambridge.

Fourie, L. 1928. "The Bushmen of South West Africa." In *The Native Tribes of South West Africa*. Cape Times, Cape Town.

Frake, C. 1962. "Cultural Ecology and Ethnography." *American Anthropologist* 64:53–59.

Fried, M. H. 1967. *The Evolution of Political Society*. Random House, New York.

Galton, F. 1889. *Narrative of an Explorer in Tropical South Africa*. Ward, Lock, London.

Geertz, C. 1966. "Religion as a Cultural System." In Banton, M. (ed.). *Anthro-*

pological Approaches to the Study of Religion. ASA Monographs 3, Tavistock, London.

Gillett, S. 1970. "Notes on the Settlement in the Ghanzi district." *Botswana Notes and Records* 2:52–55, Gaborone.

Goffman, E. 1959. *The Presentation of Self in Everyday Life.* Doubleday (Anchor Books) New York.

Gould, R. A. 1969. *Yiwara: Foragers of the Australian Desert.* Collins, London.

Guenther, M. G. 1973. "Farm Bushmen and Mission Bushmen: Social Change in a Setting of Conflicts and Pluralism of the San of the Ghanzi District, Republic of Botswana." Ph.D. thesis, University of Toronto.

Guggisberg, C. A. W. 1961. *Simba.* Timmins, Cape Town.

Gusinde, M. 1966. *Von Gelben und Schwarzen Buschmänner.* Akademische Druck und Verlagsanstalt Graz.

Hahn, T. 1881. *Tsuni-ǁgoam, the Supreme Being of the Khoi-Khoi.* Trubner, London.

Hailey, Lord. 1953. *Native Administration in the British African Territories,* Part V. The High Commission Territories: Basutoland, The Bechuanaland Protectorate, and Swaziland, HMSO, London.

Harris, M. 1968. *The Rise of Anthropological Theory.* Routledge and Kegan Paul, London.

Hawley, A. H. 1968. *Roderick D. McKenzie on Human Ecology.* University of Chicago Press. Chicago.

Heinz, H-J 1959. "The Parasitological Investigation." In Tobias, P. V. (ed.). *Provisional Report on Nuffield-Witwatersrand University Research Expedition to Kalahari Bushmen June–July, 1959.* Department of Anatomy, University of the Witwatersrand, Johannesburg.

Heinz, H-J 1966. "The Social Organization of the !Ko Bushmen." M. A. thesis, University of South Africa, Pretoria.

Heise, G. A., and Miller, G. A. 1961. "Problem Solving by Small Groups Using Various Communication Nets." In Hare, P., Borgatta, E. F., and Bales, R. F. (eds.). *Small Groups.* Knopf, New York.

Henning, M. W. 1956. *Animal Diseases in South Africa,* 3rd ed. CNA, Johannesburg.

Holub, E. 1881. *Seven Years in South Africa.* Translated by Ellen E. Frewer. 2 vols. Low, Marston, Searle and Rivington, London.

Hyde, L. W. 1971. "Groundwater Supplies in the Kalahari Area, Botswana." *Botswana Notes and Records, Special Edition,* No. 1, pp. 77–87, Gaborone.

Iberall, A. S. 1972. *Toward a General Science of Viable Systems.* McGraw-Hill, New York.

Inskeep, R. R. 1978. "The Bushmen in Prehistory." In Tobias, P. V. (ed.). *The Bushmen.* Human and Rousseau, Cape Town.

Jeffares, J. L. S. 1932. *Report on Rhodesia-Walvis Bay Reconnaissance Survey.* Government Printer, Salisbury, Rhodesia.

Katz, D., and Kahn, R. L. 1969. "Common Characteristics of Open Systems." In Emery, F. E. (ed.). *Systems Thinking,* pp. 86–104. Penguin, Ringwood, Australia.

Bibliography

Kemp, W. B. 1971. "The Flow of Energy in a Hunting Society." *Scientific American* 225, No. 3:105–15.

Kendeigh, S. C. 1961. *Animal Ecology*. Prentice-Hall, Englewood Cliffs, N. J.

Kershaw, K. A. 1964. *Quantitative and Dynamic Ecology*. Arnold, London.

King, L. C. 1963. *South African Scenery: A Textbook of Geomorphology*, 3rd ed. Oliver and Boyd, Edinburgh.

Koehler, O. 1962. "Studien zum Genussystem und Verbalbau der zentralen Khoisan-Sprachen." *Anthropos* 57:529–46.

Kokot, D. F. 1948. *An Investigation into the Evidence Bearing on Recent Climatic Changes over Southern Africa*. Irrigation Department, Government Printer, Pretoria.

Kormondy, E. J. 1969. *Concepts of Ecology*. Prentice-Hall, Englewood Cliffs, N.J.

Kremyanskiy, V. I. 1969. "Certain Peculiarities of Organisms as a 'System' from the Point of View of Physics, Cybernetics and Biology." In Emery, F. E. (ed.). *Systems Thinking*, pp. 125–46. Penguin, Ringwood, Australia.

Ladell, W. S. S. 1957. "The Influence of Environment in Arid Regions on the Biology of Man." In UNESCO. *Human and Animal Ecology*. Paris.

Lancaster, I. N. 1974. "Pans of the Southern Kalahari." *Botswana Notes and Records* 6:157–69.

Langer, W. L. 1971. "The Black Death." In Ehrlich, P. R., Holdren, J. P., and Holm, R. W. (eds.), *Man and the Ecosphere*. Freeman, San Francisco, pp. 32–7.

Laughlin, W. S. 1962. "Acquisition of Anatomical Knowledge by Ancient Man." In Washburn, S. L. (ed.). *Social Life of Early Man*. Methuen, London.

Leach, E. 1967. "Genesis as Myth." In Middleton, J. (ed.). *Myth and Cosmos*. American Museum of Natural History, Natural History Press, New York.

Lee, R. B. 1965. "Subsistence Ecology of the !Kung Bushmen." Ph.D. thesis, University of California, Berkeley.

Lee, R. B. 1968. "What Hunters Do for a Living, or, How to Make Out on Scarce Resources." In Lee, R. B., and DeVore, I. (eds.). *Man the Hunter*. Aldine, Chicago.

Lee, R. B. 1969. "!Kung Bushman Subsistence: An Input-Output Analysis." In Damas, D. (ed.). *Contributions to Anthropology: Ecological Essays*. Bulletin 230, National Museums of Canada, Ottawa.

Lee, R. B. 1972. "The !Kung Bushmen of Botswana," In Bicchieri, M. G. (ed.). *Hunters and Gatherers Today*. Holt, Rinehart and Winston, New York.

Lee, R. B. 1976. "!Kung Spatial Organisation." In Lee, R. B., and DeVore, I. (eds.). *Kalahari Hunter-Gatherers*, pp. 74–97. Harvard University Press, Cambridge, Mass.

Lee, R. B. 1979. *The !Kung San: Men, Women, and Work in a Foraging Society*. Cambridge University Press, Cambridge.

Lee, R. B., and Devore, I. 1968. "Problems in the Study of Hunters and Gatherers." In Lee, R. B., and DeVore, I. (eds.). *Man the Hunter*. Aldine, Chicago.

Bibliography

Leistner, O. A. 1967. *The Plant Ecology of the Southern Kalahari*. Botanical Survey of South Africa, Memoir No. 38. Government Printer, Pretoria.

Lenneberg, E. H. 1953. "Cognition in Ethnolinguistics." *Language,* 29:463–71.

Letty, C. 1962. *Wild Flowers of the Transvaal*. Wild Flowers of the Transvaal Book Fund, Johannesburg.

Levi-Strauss, C. 1966. *The Savage Mind*. Weidenfeld and Nicholson, London.

Levi-Strauss, C. 1969. *The Elementary Structures of Kinship*. Eyre and Spottiswoode, London.

Lichtenstein, W. H. C. 1973. *Foundation of the Cape: About the Bechuanas*. Translated and edited by Dr. O. H. Spohr. Balkema, Cape Town.

Livingstone, D. 1857. *Missionary Travels and Researches in South Africa*. Murray, London.

Lugard, F. D. 1896. "Evidence re Botswana Title to Ghanzie" (sic) 12.11.1896 Annexure 2 to Report 3 (Secretary of State's Despatch No. 108 of 23.3.1897). Botswana National Archives HC 111.

McIrvine, E. C. 1971. "Energy and Information." *Scientific American* 225, No. 3:179–88.

Mackenzie, J. 1871. *Ten Years North of the Orange River*. Edmonston and Douglas, Edinburgh.

Mackenzie, J. 1975. *Papers of John Mackenzie*. Edited by Anthony J. Dachs. Witwatersrand University Press, Johannesburg.

McLachlan, G. R., and Liversidge, R. 1957. *Roberts' Birds of South Africa*. CNA, Johannesburg.

Macworth-Praed, C. W., and Grant, C. H. B. 1963. *Birds of the Southern Third of Africa*, Vols. 1 and 2. Longmans Green, London.

Makin, W. J. 1929. *Across the Kalahari Desert*. Arrowsmith, London.

Malinowski, B. 1923. "The Problem of Meaning in Primitive Languages." Supplement 1 to Ogden, C. K., and Richards, I. A. *The Meaning of Meaning*. Kegan Paul, London.

Malinowski, B. 1935. *Coral Gardens and Their Magic*. 2 vols. Allen and Unwin, London.

Margalef, R. 1968. *Perspectives in Ecological Theory*. University of Chicago Press, Chicago.

Mark, A. K. 1971. *Cultural Ecology: The Study of Ecosystems: An Initial Formulation of a Paradigmatic Solution by Refining the Concepts of Adaptation and Regularity*. Presented to Australian and New Zealand Association for the Advancement of Science., Brisbane, May 26, 1971. (Unpublished.)

Marshall, L. 1957a. "The Kin Terminology System of the !Kung Bushmen." *Africa* 27, No. 1:1–25.

Marshall, L. 1957b. "N/ow." *Africa* 27, No. 3:232–40.

Marshall, L. 1959. "Marriage Among !Kung Bushmen." *Africa* 29, No. 4:335–65.

Marshall, L. 1960. "!Kung Bushman Bands." *Africa* 30, No. 4:325–55.

Marshall, L. 1961. "Sharing, Talking, and Giving: Relief of Social Tensions Among !Kung Bushmen." *Africa* 31, No. 3:231–49.

Bibliography

Marshall, L. 1962. "!Kung Bushman Religious Beliefs." *Africa*, 32, No. 3:221–52.

Marshall, L. 1975. "Two Ju/wa Constellations." *Botswana Notes and Records* 7:153–59, Gaborone.

Marshall-Thomas, E. 1959. *The Harmless People*. Secker and Warburg, London.

Martin, C. 1978. *Keepers of the Game: Indian Animal Relationships and the Fur Trade*. University of California Press, Berkeley.

The Masarwa (Bushmen) Report of an Enquiry by the South African District Committee of the London Missionary Society. 1935. Lovedale Press, Lovedale, South Africa.

Mauss, M. 1905. "Essai sur les variations saisonnières des sociétés Eskimo: étude de morphologie sociale." *Année Sociologique*, pp. 32–132.

Mauss, M. 1954. *The Gift: Forms and Functions of Exchange in Archaic Societies*. The Free Press, Glencoe, Ill.

Meredith, D. (hon. ed.). 1955. *The Grasses and Pastures of South Africa*. CNA, Johannesburg.

Miller, O. B. 1952. *The Woody Plants of the Bechuanaland Protectorate*. National Botanic Gardens, Kirstenbosch.

Moffat, R. 1842. *Missionary Labours and Scenes in Southern Africa*. Snow, London.

Monk, W. 1860. *Dr. Livingstone's Cambridge Lectures*. Deighton, Bell, Cambridge.

Murdock, G. P. 1949. *Social Structure*. Macmillan, New York.

Nash, M. 1966. *Primitive and Peasant Economic Systems*. Chandler, San Francisco.

Needham, R. 1971. "Remarks on the Analysis of Kinship and Marriage." In Needham, R. (ed.). *Rethinking Kinship and Marriage*. ASA Monographs 11. Tavistock, London.

Nettelton, G. E. 1934. "History of the Ngamiland Tribes up to 1926." *Bantu Studies* 8, No. 4:343–60.

Nurse, G. T., and Jenkins, T. 1977. *Health and the Hunter-Gatherer*. Karger, Basel.

Odum, E. P. 1959. *Fundamentals of Ecology*. 2nd ed. Saunders, Philadelphia.

Odum, E. P. 1971. *Fundamentals of Ecology*, 3rd ed. Saunders, Philadelphia.

Odum, H. T. 1971. *Environment, Power and Society*. Wiley (Interscience), New York.

Palmer, E., and Pitman, N. 1961. *Trees of South Africa*. Balkema, Cape Town.

Partridge, P. H. 1971. *Consent and Consensus*. Macmillan, London.

Passarge, S. 1904. *Die Kalahari*. Reimer, Berlin.

Passarge, S. 1907. *Die Buschmänner der Kalahari*. Reimer, Berlin.

Pearsall, W. H. 1962. "The Conservation of African Plains Game as a Form of Land Use." In Le Cren, E. D., and Holdgate, M. W. (eds.). *The Exploitation of Natural Animal Populations*. Blackwell, Oxford.

Peterson, N. 1976. "The natural and cultural areas of Aboriginal Australia." In Peterson, N. (ed.). *Tribes and Boundaries in Australia*. Australian Institute of Aboriginal Studies, Canberra.

Bibliography

Pettman, C. 1913. *Africanderisms: A Glossary of South African Colloquial Words and Phrases and of Place and Other Names.* Longmans Green, London.

Phillips, E. P. 1951. *The Genera of South African Flowering Plants,* 2nd ed. Department of Agriculture, Government Printer, Pretoria.

Phillipson, J. 1966. *Ecological Energetics.* Arnold, London.

Pike, J. G. 1971a. *Rainfall and Evaporation in Botswana.* U.N. Development Programme, Food and Agriculture Organization of the United Nations, Gaborone.

Pike, J. G. 1971b. "Rainfall Over Botswana." *Botswana Notes and Records, Special Edition,* No. 1, pp. 69–76, Gaborone.

Pole-Evans, I. B. 1948. *A Reconnaissance Trip Through the Eastern Portion of the Bechuanaland Protectorate and an Expedition to Ngamiland.* Department of Agriculture, Government Printer, Pretoria.

Prothero, R. M. (ed.). 1972. *People and Land in Africa South of the Sahara.* Oxford University Press, London.

Quinn, J. A. 1940. "Topical Summary of Current Literature on Human Ecology." *American Journal of Sociology,* 46 (September):191–226.

Radcliffe-Brown, A. R. 1930–31. "The Social Organization of Australian Tribes." *Oceania* 1:34–63, 206–46, 322–41, 426–56.

Radcliffe-Brown, A. R. 1950. "Introduction." In Radcliffe-Brown, A. R., and Forde, D. (eds.). *African Systems of Kinship and Marriage.* Oxford University Press, London.

Rappaport, R. A. 1971. "The Flow of Energy in an Agricultural Society," *Scientific American* 225, No. 3:116–32.

Revelle, R., and Landsberg, H. H. (eds.). 1970. *America's Changing Environment.* Beacon Press, Boston.

Reynolds, G. W. 1950. *The Aloes of South Africa.* The Aloes of South Africa Book Fund, Johannesburg.

Rheinallt-Jones, J. D., and Doke, C. M. (eds.). 1937. *Bushmen of the Southern Kalahari.* Witwatersrand University Press, Johannesburg.

Roberts, A. 1951. *The Mammals of South Africa.* CNA, Johannesburg.

Rose Innes, R. 1964. "Discontinuous distribution of the gerbil flea, *Xenopsylla philoxera* in Southern Africa." In Davis, D. H. S. (ed.). *Ecological Studies in Southern Africa.* Junk, The Hague.

Sahlins, M. 1965. "On the Sociology of Primitive Exchange." In *The Relevance of Models for Social Anthropology.* ASA Monographs 1. Tavistock, London.

Salzano, F. M. 1972. "Genetic Aspects of the Demography of American Indians and Eskimos." In Harrison, G. A., and Boyce, A. J. (eds.). *The Structure of Human Populations.* Oxford University Press, London.

Sapir, E. 1966. "Conceptual Categories in Primitive Languages." In Hymes, D. (ed.). *Language in Culture and Society.* Harper and Row, New York.

Schaller, G. B. 1972. *The Serengeti Lion.* University of Chicago Press, Chicago.

Schapera, I. 1930. *The Khoisan Peoples of South Africa.* Routledge and Kegan Paul, London.

Schapera, I. 1939. "A Survey of the Bushman Question." In *Race Relations* 6, no. 2:68–83.

313

Bibliography

Schinz, H. 1891. *Deutsch–Südwest-Afrika*. N.p., Oldenburg and Leipzig.
Schmidt-Nielsen, K. 1964. *Desert Animals*. Oxford University Press, London.
Schulze, L. 1907. *Aus Namaland und Kalahari*. Fischer, Jena.
Schumann, T. E. W. 1941. *Atmospheric Pressures and Weather Charts*. Union Meteorological Office, Pretoria.
Schwarz, E. H. L. 1928. *The Kalahari and Its Native Races*. Witherby, London.
Service, E. R. 1971. *Primitive Social Organization*. Random House, New York.
Shaw, E. M., Woolley, P. L., and Rae, F. A. 1963. "Bushman Arrow Poisons." *Cimbebasia*, No. 7:2–41, Windhoek.
Silberbauer, G. B. 1956. *Grazing Survey: Ngamiland*. Report to the Bechuanaland Protectorate Government, Mateking.
Silberbauer, G. B. 1963. "Marriage and the Girl's Puberty Ceremony of the G/wi Bushmen." *Africa* 33, No. 3:12–24.
Silberbauer, G. B. 1965. *Bushman Survey Report*. Bechuanaland Government, Gaberones.
Silberbauer, G. B. 1971. "Ecology of the Ernabella Aboriginal Community." *Anthropological Forum* 3, No. 1:21–36.
Silberbauer, G. B. 1972. "The G/wi Bushmen." In Bicchieri, M. G. (ed.). *Hunters and Gatherers Today*. Holt, Rinehart and Winston, New York.
Silberbauer, G. B., and Kuper, A. 1966. "Kgalagari Masters and Bushman Serfs." *African Studies* 25, No. 4:171–79.
Singer, R. 1978. "The Biology of the San." In Tobias, P. V. (ed.). *The Bushmen*. Human and Rousseau, Cape Town.
Skaife, S. H. 1953. *African Insect Life*. Longmans Green, London.
Smith, M. G. 1955. *The Economy of Hausa Communities of Zaria*. HMSO, London.
Smith, M. G. 1960. *Government of Zazzau 1800–1950*. Oxford University Press, London.
Smithers, Reay H. N. 1964. *A Checklist of the Birds of the Bechuanaland Protectorate and the Caprivi Strip*. National Museums of Southern Rhodesia, Salisbury.
Smithers, Reay H. N. 1971. *The Mammals of Botswana*. Museum Memoir No. 4. The Trustees of the National Museums of Rhodesia, Salisbury.
Southall, A. 1965. "A Critique of the Typology of States and Political Systems." In *Political Systems and the Distribution of Power*. ASA Monographs 2. Tavistock, London.
Spedding, C. R. W. 1971. *Grassland Ecology*. Oxford University Press, London.
Stapleton, C. C. 1955. *Common Transvaal Trees*, 2nd ed. Department of Forestry, Government Printer, Pretoria.
Stengel, H. W. 1963. "Evaporation and Sedimentation in Storage Dams." In Stengel, H. W. (ed.). *Water Affairs in S. W. A.* Afrika-Verlag der Kreis, Windhoek.
Steward, J. H. 1955. *Theory of Cultural Change*. University of Illinois Press, Urbana.
Story, R. 1958. *Some Plants Used by the Bushmen in Obtaining Food and Water*. Department of Agriculture, Government Printer, Pretoria.

Bibliography

Story, R. 1964. "Plant Lore of the Bushmen." In Davis, D. H. S. (ed.). *Ecological Studies in Southern Africa*. Junk, The Hague.

Stott, D. H. 1962. "Cultural and Natural Checks on Population Growth." In Montagu, M. F. A. (ed.). *Culture and the Evolution of Man*. Oxford University Press, London.

Stow, G. 1905. *The Native Races of South Africa*. Swan Sonnenschein, London.

Tabler, E. C. 1955. *The Far Interior*. Balkema, Cape Town.

Tagart, E. S. B. 1931. *Report on the Masarwa and on Corporal Punishment Among Natives in the Bamangwato Reserve of the Bechuanaland Protectorate*. Dominions No. 136, Confidential, October.

Tanaka, J. 1969. "The Ecology and Social Structure of Central Kalahari Bushmen: A Preliminary Report." *Kyoto University African Studies* 3:1–26.

Tanaka, J. 1976. "Subsistence Ecology of Central Kalahari San." In Lee, R. B., and DeVore, I. (eds.). *Kalahari Hunter-Gatherers*. Harvard University Press, Cambridge, Mass.

Thomas, W. L. (ed.). 1956. *Man's Role in Changing the Face of the Earth*. 2 vols. University of Chicago Press, Chicago.

Thomson, A. L. 1964. *A New Dictionary of Birds*. British Ornithologists' Union. Nelson, London.

Tobias, P. V. 1956. "On the Survival of the Bushmen." *Africa* 26, No. 2 (April):174–86.

Tobias, P. V. 1957. "Bushmen of the Kalahari." *Man* 36 (March):1–8.

Tobias, P. V. 1964. "Bushman hunter-gatherers: A study in human ecology." In Davis, D. H. S. (ed.). *Ecological Studies in Southern Africa*. Junk, The Hague.

Tobias, P. V. 1978. "Introduction to the Bushmen or San." In Tobias, P. V. (ed.). *The Bushmen*. Human and Rousseau, Cape Town.

Traill, A. 1978. "The Languages of the Bushmen." In Tobias, P. V. (ed.). *The Bushmen*. Human and Rousseau, Cape Town.

Turnbull, C. M. 1968. "The Importance of Flux in Two Hunting Societies." In Lee, R. B., and DeVore, I. (eds.). *Man the Hunter*. Aldine, Chicago.

Turnbull, C. M. 1972. "Demography of Small-scale Societies." In Harrison, G. A., and Boyce, A. J. (eds.). *The Structure of Human Populations*. Oxford University Press, London.

Turnbull, C. M. 1974. *The Mountain People*. Picador, Pan Books, London.

van Straten, O. J. 1963. "A note on the ground water potential of certain areas adjacent to the main internal drainage system of the northern Bechuanaland Protectorate." In Langdale-Brown, I., and Spooner, R. J., (eds.). *Land Use Prospects of Northern Bechuanaland*, Appendix II. Directorate of Overseas Surveys, Tolworth, Surrey.

Vayda, A. P., and Rappaport, R. A. 1968. "Ecology, Cultural and Non-cultural." In Clifton, J. A. (ed.). *Introduction to Cultural Anthropology*. Houghton Mifflin, Boston.

von Bertalanffy, L. 1968. *General System Theory*. Braziller, New York.

von Breitenbach, F. 1965. *The Indigenous Trees of Southern Africa*. Government Printer, Pretoria.

von La Chevallerie, M. 1970. "Meat Production from Wild Ungulates." *Pro-*

ceedings of the South African Society of Animal Production 9:73–87, Pretoria.

Wallace, A. F. C. 1961. *Culture and Personality*. Random House, New York.

Watt, J. M., and Breyer-Brandwijk, M. G. 1962. *The Medicinal and Poisonous Plants of Southern and Eastern Africa*. Livingstone, Edinburgh.

Weare, P. R., and Yalala, A. 1971. "Provisional Vegetation Map of Botswana." *Botswana Notes and Records* 3:131–48, Gaborone.

Weber, M. 1966. *The Theory of Social and Economic Organisation*. The Free Press, Glencoe, Ill.

Wellington, J. 1955. *Southern Africa: a Geographical Study*. 2 vols. Cambridge University Press, Cambridge.

Wellington, J. 1967. *South West Africa and Its Human Issues*. Oxford University Press, London.

Welty, J. C. 1964. *The Life of Birds*. Constable, London.

West, O. 1955. "Veld Management in the Dry, Summer-Rainfall Bushveld." In Meredith, D. (ed.). *The Grasses and Pastures of South Africa*. CNA, Johannesburg.

Westphal, E. O. J. 1963. "The Linguistic Prehistory of Southern Africa: Bush, Kwadi, Hottentot and Bantu Linguistic Relationships." *Africa* 33, No. 3:237–65.

White, A., and Sloane, B. L. 1937. *The Stapelieae*, 3 vols. Haselton, Pasadena, Calif.

White, L. A. 1943. "Energy and the Evolution of Culture." *American Anthropologist* 35, No. 3, pt. 1:335–56.

Whiteman, P. T. S. 1971. "Limitations to Crop Production in the Kalahari." *Botswana Notes and Records, Special Edition*, No. 1, pp. 114–21, Gaborone.

Whittaker, R. H. 1970. *Communities and Ecosystems*. Macmillan, London.

Whittingham, P. 1965. "Problems of Survival." In Edholm, O. G., and Bacharach, A. L. (eds.). *Exploration Medicine*. Wright, Bristol.

Willis, E. O. 1967. *The Behavior of Bicolored Antbirds*. University of California Publications in Zoology, vol. 79. University of California Press, Berkeley.

Wright, J. B. 1971. *Bushman Raiders of the Drakensberg 1840–1870*. University of Natal Press, Natal.

Wyndham, C. H. 1956. "Heat Regulation of Ma Sarwa (Bushmen)." *Nature* 178 (October 20):869–70.

Wyndham, C. H., Strydom, N. B., Ward, J. S., Morrison, J. F., Williams, C. G., Bredell, G. A. G., von Rahden, M. J. E., Holdsworth, L. D., van Graan, C. H., van Rensburg, A. J., and Munro, A. 1964. "Physiological reactions to heat of Bushman and of unacclimatized and acclimatized Bantu." *Journal of Applied Physiology* 16, No. 5 (September):885–8.

Yellen, J. E., and Lee, R. B. 1976. "The Dobe-/Du/da environment: Background to a hunting and gathering way of life." In Lee, R. B., and DeVore, I. (eds.). *Kalahari Hunter-Gatherers*. Harvard University Press, Cambridge. Mass.

Bibliography

Yengoyan, A. A. 1966. "Ecological Analysis and Agriculture." *Comparative Studies in Society and History* 9, No. 1 (October):105–17.

Yengoyan, A. A. 1968. "Demographic and Ecological Influences on Aborignal Marriage Sections." In Lee, R. B., and DeVore, I. (eds.). *Man the Hunter.* Aldine, Chicago.

Index

319

Index